Climate Change
and
British Wildlife

Climate Change and British Wildlife

Trevor Beebee

BLOOMSBURY WILDLIFE
LONDON · OXFORD · NEW YORK · NEW DELHI · SYDNEY

Dedication

To Maggie, lifelong companion in wild places; and to Malcolm Smith and John Clegg, inspired naturalists of an enviable generation.

Half-title: Snowdrops in melting snow.
Frontispiece: Large waves engulfing a local train at Dawlish on the south coast of Devon during a storm.

BLOOMSBURY WILDLIFE
Bloomsbury Publishing Plc
50 Bedford Square, London, WC1B 3DP, UK

BLOOMSBURY, BLOOMSBURY WILDLIFE and the Diana logo are trademarks of
Bloomsbury Publishing Plc

First published in Great Britain 2018

Copyright © Trevor Beebee, 2018

Trevor Beebee has asserted his right under the Copyright, Designs and Patents Act, 1988, to be identified as Author of this work

For legal purposes the acknowledgements on p. 352–3 constitute an extension of this copyright page

All rights reserved. No part of this publication may be reproduced or transmitted in any form or by any means, electronic or mechanical, including photocopying, recording, or any information storage or retrieval system, without prior permission in writing from the publishers

Bloomsbury Publishing Plc does not have any control over, or responsibility for, any third-party websites referred to or in this book. All internet addresses given in this book were correct at the time of going to press. The author and publisher regret any inconvenience caused if addresses have changed or sites have ceased to exist, but can accept no responsibility for any such changes

A catalogue record for this book is available from the British Library

ISBN: HB: 978-1-4729-4320-0; ePDF: 978-1-4729-4317-0; ePub: 978-1-4729-4319-4

2 4 6 8 10 9 7 5 3 1

Page layouts by Susan McIntyre
Jacket artwork by Carry Akroyd
Printed and bound in China by C&C Offset Printing Co., Ltd

To find out more about our authors and books visit www.bloomsbury.com and sign up for our newsletters

Contents

Preface	6
1 What's going on?	10
2 How are plants responding?	38
3 Invertebrate tales	74
4 Freshwater and terrestrial vertebrates	118
5 Fungi, lichens and microbes	162
6 Freshwater and terrestrial communities	180
7 Coastal and marine environments	214
8 As time goes by	244
9 What the future may hold	266
10 Conservation in a warming world	306
References	338
Abbreviations	344
Species names	345
Credits	352
Index	354

Preface

A hazy sun climbs hesitantly above Glastonbury Tor, while Garden Bumblebees drone peacefully among the creamy blossoms of a sprawling honeysuckle. It's already shirt-sleeve weather on one of those perfect, spirit-raising days that announce the start of another spring. After dusk, Smooth and Palmate Newts tread warily across the floor of a garden pond. But this is not spring, at least not by any calendar definition. It is New Year's Day 2016 and the scene is not unusual. For years now, bumblebees have appeared every month, indeed on most days, in our garden from September through to April. Newts have been in the pond since November, and by the end of December the first wild garlic flowers are in bloom, trailblazers of what will shortly become a glorious white carpet. Occasional daffodils, normally expected for March's Mothering Sunday, have also raised their yellow flags above the parapet.

It hasn't always been like this. Something is afoot, the seasons are changing and naturalists (at least, those a bit long in the tooth) have been among the first to notice it. Even so, the realisation came slowly because no two years are ever the same and there have always been unusually early springs and mild winters. On 16 January 1777, Gilbert White recorded in his diary: 'bees and flies moving, air full of insects; spiders shoot their webs: butterfly out'. But what White described much more often in *The Natural History of Selborne* were winter conditions that are now rarely or never seen in southern England:

> ... *meat frozen so hard it can't be spitted; several of the thrush kind are frozen to death; the Thames, it seems, is so frozen that fairs have been kept on it; rugged, Siberian weather; many rooks, which attempting to fly fell from the trees with their wings frozen together by the sleet, that froze as it fell ...*

and so on. Recognising true trends in climate involves a statistical element, separating real, long-term change from the background noise of occasional, exceptional years. This separation has now been accomplished, and significant changes in the British climate over the past few decades have been demonstrated beyond reasonable doubt.

In this book I briefly review the evidence for climate change, especially trends in temperature and rainfall in the United Kingdom over the past 50 or so years, and ways in which these changes could affect our wildlife. The bulk of the text, however, describes the actual, reported changes across a wide range of taxonomic groups including plant, invertebrate, fungal and vertebrate species, together with impacts at the level of ecological communities. Whenever available, these changes refer to the results of serious scientific studies, but I make no apology for also including occasional less robust inferences, essentially anecdotal, where they seem credible. However, I clearly distinguish this kind of evidence wherever I allude to it. For each species I give both the common and scientific names on first mention but thereafter only the common ones where such exist, except in tables where scientific names are always included. At the end of each chapter I discuss whether it is possible to draw any generalisations about climate change effects across the various taxonomic groups.

British Wildlife magazine has published many articles about the impacts of climate change, the first very shortly after its inauguration in 1989, and no doubt many naturalists are broadly aware of what has been going on. I hope, nevertheless, to interest both those already familiar with the subject and others looking for an overview by drawing together a wide range of material in this single volume. Although there is now a plethora of books on actual and potential impacts of climate change, most are concerned with broad issues such as environmental consequences, economic impacts and other potential disasters for humans. Precious few major on wildlife, and, at the time of writing, none has attempted to assemble evidence for what is happening to plants, fungi and animals in the UK. Thus Schneider & Root (2002) listed some North American case studies, Brodie *et al.* (2012) reported effects on vertebrates worldwide, Pearce-Higgins & Green (2014) were concerned with birds, while Root *et al.* (2015) focused on evidence in California. All good reads, but excepting Pearce-Higgins & Green, hardly any mention of Europe, let alone the UK. Yet we have the longest history of wildlife recording anywhere in the world and are therefore in pole position for studying how climate has influenced our flora and fauna over at least decades, and in some cases centuries.

I should make clear what this book does not intend. I have not attempted to make it into a scientific review of the standard type, punctuated with references that try and cover every eventuality. There are references – but, for the most part, I have limited them to just one or two for each topic area. Another ambition was to avoid making the book a simple list of all the plants, fungi and animals that have been affected by climate change. There are, of course, plenty of examples, but I have tried to write them into explanations of the whys and wherefores of how our wildlife is responding to a warming countryside. Then there is a chapter devoted to people's perceptions of climate change in the UK, including some views of scientists and politicians. While focusing on events for which there is already evidence of change, I could not, and would not wish to, ignore altogether the many prognostications of what we might be in for in the coming years. These predictions test human ingenuity to its limits of credibility, and perhaps beyond, and I have confined such considerations to the penultimate chapter. In the final chapter I contemplate the implications of climate change for wildlife conservation, ironically including the dangers posed by human activities devoted to countering climate change effects.

My natural habitat is mostly wetland, the places where amphibians and aquatic insects lurk among wafting waterweeds. These enchanting corners of the countryside entrapped me early in life, mostly thanks to a nearby farm pond long since obliterated under concrete. For me, as for many friends and colleagues, the realisation that climate change was impacting our wildlife dawned slowly, and mostly as a result of happenings in my back garden. Probing more widely into the subject for this book has been a real eye-opener, a mixture of both delight at new arrivals and serious concern about species declines. I'm still not sure what will be the final, dominant emotion. The wheel is very much still in spin.

I have many people to thank for assistance with the writing of this tome. Katy Roper, of Bloomsbury Publishing, could not have been more supportive from start to finish. I am also grateful to Hugh Brazier for meticulously improving bits of the text, and for spotting occasional silly mistakes, and to James Pearce-Higgins for reviewing and making helpful comments on the text. I am indebted, too, to friends and fellow travellers who have volunteered their personal experiences (including quotations) of how climate change has impacted on the species and places they know so well, and pointed me in directions of enquiry I might well have otherwise missed; unless stated otherwise, quotations in the text are comments sent directly to me by the person named. So thank you John Altringham, Brian Banks, John Barkham, Ray Barnett, Katherine Boughey, Steve Brooks, John F. Burton, David Cabot, Harry Clarke, Sandy Coppins, Keith Corbett, Genevieve Dalley, Jonty Denton, Chris Ellis, Sue Everett, Garth Foster, Tim Gardiner, Chris Gleed-Owen, Mark Hill, Mike Hutchings, Jeremy Kerr, John Knowler, Dafydd Lewis, Nick Littlewood, Craig Macadam, Richard Mabey, Peter Marren, Charlotte Marshall, Brian Moss, Geoff Oxford, Chris Packham, James Pearce-Higgins, Chris Preston, Tim Sparks, Alan Stewart, Chris Thomas, Peter Topley, Kevin Walker and Jon Webster. Your contributions made my task easier than it would otherwise have been. It goes without saying that any errors in this treatise are mine and mine alone.

In addition, my gratitude is heartfelt to contributors who have offered images free of charge for use in this book. They are Jim Barton, Paul Cecil, Jonty Denton, Lars Eklundh, Chris Ellis, Caroline Fitton, Tim Gardiner, Chris Gleed-Owen, David Goddard, Hongxiao Jin, Roger Key, Leanne Manchester of the Wildlife Trusts, Nick Owens, James Pearce-Higgins, Jill Pelto, Mike Pennington, George Peterken, Jean-Yves Sgro, Graham Shephard of the Rothamsted Institute, Tim Sparks, Charlotte Sullivan, Chris Thomas and Nigel Voaden. Special thanks also to Kathy Jetñil-Kijiner for permission to reproduce part of her poem 'Utilomar' and to Rachel McCarthy for permission to reproduce the entire poem 'Survey North of 60 degrees'. Your generous input is much appreciated.

Finally, thanks to Carry Akroyd, whose covers are always a wonder to behold.

Trevor Beebee, January 2018

What's going on?

chapter one

Climate change is not a new phenomenon. On any timescale longer than a few human generations, climatic variation has been commonplace for as far back as we can investigate, based on a range of indirect measures such as thicknesses of tree rings, oxygen isotopes in ice cores, fossil species distributions and many other lines of evidence. Within our three score years and ten, most folk in temperate latitudes become used to sporadic variations – warm or cool summers, cold or mild winters – without any expectation of permanent change one way or the other. The bigger picture is very different from what most of us perceive as the day-to-day reality.

Evidence for change

Over the last million years Britain has been in the grip of intermittent glaciations, including a spell between 22,000 and 20,000 years ago when temperatures were so low that no humans could survive anywhere in the British Isles. Our ancestors cut their teeth in that ferocious winter, wandering for generations in the shadow of the ice fields but leaving no written record of their passing. In the words of the 20th-century American science writer Loren Eiseley,

> With our short memory, we accept the present climate as normal. It is as though a man with a huge volume of a thousand pages before him ... should read the final sentence on the last page and pronounce it history.

Climatic oscillations during the Pleistocene ice ages were enormous. Time after time the great glaciers crept forth, grinding all before them, but then hesitating and retreating as if to inspect a devastated landscape before their inevitable return. During one such interglacial armistice, 120,000 years ago, hippos, lions and hyenas roamed a subtropical Thames valley.

OPPOSITE PAGE:
'Time for Action' climate change demonstration in London, March 2015.

These dramatic transitions were not always slow to take effect. At the start of the Holocene, the interglacial warm period we currently enjoy, temperatures rose several degrees Celsius within about 50 years while North Sea shorelines retreated by 100m annually as sea levels rose. Such huge changes must have had profound consequences for humans and wildlife alike, all within single human lifetimes. Today, with sea levels rising again, many coral islanders in tropical oceans face a rerun of the early Holocene floods – but without any convenient retreat.

Closer to the present day, the UK has experienced substantial climate variation within the last 1,000 years. It is possible to reconstruct the experiences of our immediate ancestors by assembling records from a wide range of written sources that long pre-date the arrival of scientific devices for measuring climate, as detailed by Kington (2010). A warm spell extending over five centuries, and ending around 700 years ago, permitted the growth of grape vines as far north as the Cheshire plain. Wine from Manchester! However, this was followed by the 'Little Ice Age', also extending over a few hundred years, during which temperatures plummeted and winter fairs were held on a frozen River Thames. Kington also demonstrated, however, that these warm and cold spells were far from uniform. Even during the Maunder Minimum, the coldest period of the Little Ice Age, Scotland recorded its best harvest for 60 years in 1668.

Winter scene during the 'Little Ice Age' in the 18th century, when the Thames regularly froze over in London.

The ups, downs and eccentricities of climate over short periods revealed by looking closely at historical records are a pertinent reminder that interpreting climate change needs a long-term perspective. These events must have impacted our wildlife, but in ways we will probably never know. One thing all the past changes have in common is that they cannot be blamed on humans; as far as we understand it, the key drivers were variations in the earth's orbital geometry combined with fluctuations in solar output. All of which forms the backdrop for current interest in climate change, which is now (almost) universally agreed to be the result of increased human economic activity, especially the emissions of greenhouse gases and consequent warming of the entire planet. The climate crisis is just the most recent of a dismal catalogue of human assaults on our precious life-support systems. Pesticides in the 1960s, acid rain in the 1980s and ozone holes in the 1990s all tested the very soul of scientific endeavour and did not find it wanting. Scepticism is a prerequisite for good science, but those still doubting that this round of climate change is our fault might pause to reflect on a scientific track record that has proved right every time environmental disasters have loomed. Svante Arrhenius, a Nobel Prize-winning chemist, made the connection between fossil fuels, carbon dioxide and the 'greenhouse effect' at the end of the 19th century. A hundred years on we can but admire his perspicacity.

Evidence about recent climate events is convincingly summarised in the famous 'hockey stick' graph, which shows that following many decades of approximate stability, global temperatures began to rise sharply at the start of the 20th century. After stalling briefly during the 1970s, an upward trend reasserted itself and was almost exponential over the next 30 years. Warming has continued in the 21st century, albeit more slowly, such that by 2015 the global average temperature was about 1°C higher that it was 100 years earlier. This dramatic change mirrored increases in atmospheric carbon dioxide concentrations, which in turn were generated primarily by the burning of fossil fuels. Those satanic mills of the Industrial Revolution heralded a cardinal change that has long survived the human squalor for which they are most often remembered. An interesting perspective on the 1°C increase is that global average temperatures at the height of the last glaciation were only about 5°C cooler than now. Even so, averages can be poor indicators, smothering the interesting variation from which they come. In this case the mean recent temperature increase disguises very considerable differences across the globe, with polar regions (especially the Arctic) warming substantially more, and tropical regions less, than the 1°C average.

Thanks to one of the world's most comprehensive arrays of climate recording stations, events in the United Kingdom over recent centuries are particularly well documented. For biologists concerned with climate there is a curious coincidence. Robert FitzRoy, captain of the *Beagle* and confidant of Charles Darwin on his uncharted voyage into the past, had a creative vision all of his own when he inaugurated the UK's Meteorological Department in 1854. It is largely thanks to FitzRoy and his successors that climate change in the UK can be compared with the global pattern. British data held by the Meteorological Office's Hadley Centre includes temperature records for central England (HadCET) going back more than 350 years, the longest series anywhere in the world. CET temperatures correlate strongly with those measured at all but one of 24 other recording stations distributed around the UK between 1964 and 2004. This has allowed CET records to be widely used as yardsticks in phenology (the study of the timing of natural events), because they are representative of climate trends across the entire country.

Climate change sceptics like to point out that the temperature pattern revealed by the CET records, similar to global data sets, looks more like a step change than an ongoing trend, which suggests that we may have crossed the finishing line. Could they be right? Most climate scientists don't think so, as Richard Kerr (2009) retorted in a *Science* paper when the apparent hiatus in temperature increase became evident: 'What happened to global warming? Scientists say just wait a bit.' It looks like he was right; recent trend data, up to 2015 (see figure opposite), confirm that a continuous increase is ongoing. Glaciers are still retreating, Arctic sea ice edges forth more cautiously every summer, and 2016 was the warmest year (globally) on record. Extreme weather events, widely predicted consequences of global warming, are in the news almost every month somewhere in the world.

Unpredictability in the short term is another expectation as climates change, so odd patterns that seem contradictory to overall trends should not surprise us. Three recent springs in Britain included, in March or April, extended cold periods that felt more like winter than the months preceding them. Amphibian breeding migrations were all over the place, some earlier than usual owing to exceptionally mild winters and others later, caught out by unexpected frosts at a critical time. Blasts of cool air coming our way as a result of Arctic warming

What's going on?

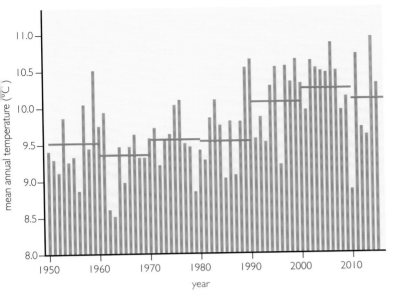

Mean annual temperatures in central England, 1950–2015. Blue bars show the averages for each decade, and for the six years 2010–2015 inclusive. Based on data sourced from the Met Office.

depressing the jet stream has been touted as an explanation. Spring is a critical time for virtually all British wildlife, and there could be long-term repercussions if cold spells at this time of year are more than a short-term blip.

There have been some differences in temperature trends between the UK and the global average, especially the lack of continuous increases during the first decades of the 20th century. It is only since the 1980s that the UK has warmed in similar fashion to the world as a whole. This makes personal sense to me, recalling the 1960s when newts mostly arrived in ponds in early spring when the textbooks told us they should. But after the early 1980s climate change in the UK mirrored what was happening across the planet. The numbers are telling. Mean average annual temperatures rose by more than 9 per cent compared with pre-1980s measurements, and rainfall also increased, though by relatively less (about 6–7 per cent overall). However, the changes have not been spread evenly across the seasons. Springs, autumns and winters have warmed by around 11–15 per cent, summers by much less, at around 5 per cent. Those hoping for Mediterranean-style holidays at home have, at least so far, been disappointed. This pattern of temperature change has been broadly similar across the UK, from southern England to Scotland, but rainfall has not been so consistent. Scotland has experienced a much larger increase (at least 16 per cent) than has happened in England

A Garden Bumblebee, now regularly seen on the wing all year round.

Flooding events in towns are increasingly regular disasters caused by increased storminess that may be a consequence of climate change.

(about 5 per cent). In England there has been little difference between rainfall changes in summer and winter, whereas in Scotland the biggest increases have been in winter. Predictions of drier summers in southern England have, at the time of writing, yet to materialise.

Naturalists are quick to notice not only the obvious effects of climate change but also the oddities and discrepancies that, like cold springs and the not obviously improved summers, give pause for thought. Jonty Denton is a naturalist with over 12,000 species on his observation list. His observations (unpublished) no doubt mirror the experiences of many others:

From 1990ish to 2005, late summers were on the whole warm and stable and quite dry. Lo and behold, species which developed then romped north: Long-winged Conehead (new for north Hampshire in 1991) and nearly in Scotland now; Roesel's Bush-cricket was the funniest, as each year in the 1990s, as you drove up into East Anglia you could wind your window down and see how far they had got up the M11! This has broken down since the mid-00ies and a wet summer trend predominates.

What's going on?

These glitches emphasise that cold statistics do not tell the whole story. Anybody living in a flood-risk zone, unfortunately an ever-increasing proportion of the British population, will be well aware that rainfall has also become increasingly episodic. Averages are again deceptive and mean little when three months' worth of expected rain falls within a couple of days. 'Storminess' has become much more common, and severe, especially since 2000. The increase in extreme weather events is important in the context of the environmental as well as the economic impacts of climate change. It is after all weather, the up close and personal manifestation of climate, that impacts most directly on wildlife as well as on people.

What might all this mean for wildlife?

All living organisms depend on favourable environments to survive, which means that every species has a range of specific needs outside of which it cannot prosper. These 'tolerance envelopes', or niches, relate to many aspects of life, only one of which is climate. However, there is a widespread consensus that 'climate envelopes' are of major, often overriding, importance in dictating species distributions. For terrestrial species, temperature, rainfall and geology are prime determinants of distribution, while in the sea we can add salinity, tides, currents and substrate, but remove rainfall. In both environments,

Wildlife enthusiasts recognise the potential threat from climate change to the UK's flora and fauna, here at a demonstration in Trafalgar Square.

The distribution of Small-leaved Lime in the UK is limited by summer temperature and may expand as a consequence of any future warming trend.

temperature stands out as the most important single factor. It is worth exploring briefly why this is.

On land and in freshwater, temperature is critical to all stages of plant development, from production and dispersal of seeds through germination to growth rate and photosynthetic activity. The Small-leaved Lime *Tilia cordata* is an illuminating example of how temperature can limit distribution. This tree occurs no further north than Cumbria. In that region seeds are rarely produced, and are infertile when they do appear. Growth of pollen tubes is highly temperature-dependent, and at the tree's range edge the tubes do not grow far enough down the style of the female flower to permit fertilisation of the ovary. Presumably the Small-leaved Lime spread to its northern outposts when the climate was warmer than it is today. Efficient seed production requires average August temperatures of at least 20°C; in Cumbria during the mid-20th century this average was a little above 18°C.

Another important and very different aspect of temperature relations applies to plants that benefit from a period of cold, known as vernalisation, as a prerequisite for the production of flowers, pollen or seeds, or for germination. The herb Cloudberry *Rubus chamaemorus*, with its succulent golden yellow fruits, is an inhabitant of high moors and blanket bogs. The seeds of this plant require at least five months with temperatures between 4°C and 5°C before

What's going on?

Many plants such as Cloudberry have distribution patterns determined by a combination of climate and other factors, such as substrate, not climate alone.

germination can occur. This physiological requirement for a period of low temperatures in winter is by no means uncommon and has equivalents in some insects and amphibians. Complex genetics that underpin vernalisation have been unravelled, but any adaptive benefits of the process remain debatable. Whatever the underlying reasons for vernalisation, a warming climate, especially in winter, will not benefit plants or animals that require it.

The interacting biochemistries of respiration and photosynthesis can also contribute to plant distribution limits. For any species, there is a narrow range of temperatures over which photosynthesis, and thus the ability to grow and prosper, dominates respiration, a process that essentially uses up resources. The upper altitudinal limit of Scots Pine *Pinus sylvestris* is lower in Scotland than in Norway, and lower on Scotland's west coast than in the east. Norway's more continental climate allows efficient photosynthesis in summer at high altitudes while cold winters minimise respiratory losses. In Scotland, wet and cloudy summers are less supportive of photosynthesis but mild winters support high respiration rates, all of which limit how far up a mountainside the tree can prosper.

The ability of Scots Pine (shown here germinating) to thrive depends on complex interactions between climate and physiology that lead to variable altitudinal limits in different places.

Mistletoe has a complex set of summer and winter temperature requirements that vary between the east and west of the UK.

With so many interacting processes, it is scarcely surprising that the climate envelopes that constrain species ranges are not always simple to deduce. Climate envelopes are defined as areas of habitat within which conditions such as temperature and rainfall are adequate to support a species. An obvious method of plotting distributions in relation to, for example, summer isotherms and looking for the best fit often fails to produce a convincing picture. The distribution of Mistletoe *Viscum album* in north-west Europe is temperature-related, but in a complex way. Where summer temperatures are high enough, as in south-east England, Mistletoe can survive winter minima as low as −8°C. In regions with cooler summers, such as in parts of the West Country, the plant's tolerance of cold winters decreases. Its distribution limits are dictated by a combination of temperature regimes at very different times of year.

Temperature has direct effects on animals too, especially on ectotherms. These are species without the ability to control their body temperatures by internal heat generation. They are sometimes, but inappropriately, called 'cold blooded'. Ectotherms rely on external sources such as sunshine or infrared radiation to vary their temperature, and in summer they can be as warm as endotherms, the mammals and birds that generate their body heat internally. Most ectotherms enter a period of dormancy during the winter months, seeking shelter from extremes of cold but varying in their susceptibility to freezing conditions. In some cases winter minimum temperatures can limit distributions, as may be the case for the Round-mouthed Snail *Pomatias elegans*. This gastropod inhabits calcareous soils of south-east England and Wales but is absent from apparently suitable habitats in the more 'continental' region of East Anglia. Even so, summer maximum temperatures are more significant than winter minima for many ectotherms. In dragonflies, temperature influences egg hatching, larval feeding rates, development time and final metamorphosis. It also determines adult flight periods and, in some species, colour change. Female Emperor Dragonflies *Anax imperator* are predominantly green in moderate summer heat but turn blue, like males, if the thermostat rises high enough. Why this happens is anybody's guess. Natterjack Toads *Bufo calamita* are at their northern range limit in the British Isles and breed most successfully when warm springs and early summers support rapid tadpole growth rates. Nevertheless, they are on a cusp: too much sunshine desiccates the temporary ponds they need for breeding, while cooler weather compromises tadpole survival.

Climate change in the UK is associated with increasing average temperatures, higher levels of rainfall and greater frequencies of storms and other extreme weather events. As we will see, so far at least it is temperature that has driven most of the wildlife responses. However, increased warmth is a double-edged sword because, while many species benefit from it, a substantial number do not. At first sight this seems odd, since extra heat speeds up growth and development, surely universal benefits for any plant or animal. Yet there are species that have opted out of this comfort zone, choosing instead mountain-tops or even Arctic tundra. Evolution has a straightforward explanation for how this came about. The best conditions for life result in an intense struggle between rival species all trying to live together. The choice becomes one of striving against a host of would-be competitors, or finding another niche altogether. Choosing cool habitats is a workable strategy because proteins, the functional components of all living cells, are astonishingly adaptable. That is why some bacteria thrive in Yellowstone's boiling springs and there are lichens in Antarctica's dry valleys. All very well, but living on the edge of what is biologically feasible comes at a cost. Any significant climatic change can tip the balance towards extinction. On the other hand, when the glaciers eventually sneak forth once again, as they inevitably will, the species of today's wind-blown highland screes and snowfields may once again rule the roost over great swathes of northern Europe.

Exploring the evidence

The impact of climate change on British wildlife deserves, and has received, serious scientific attention. Here I outline some of the methods employed to investigate what is going on. Time, then, for some hypotheses. This formality of science is important for its credibility. FitzRoy was lambasted because his weather forecasting was not based on an explicit hypothesis; no matter that he usually got it right. Normally, though, ideas have limited value unless they can be rigorously tested, a concept from which theoretical physicists engaged in cosmology seem increasingly distanced. Fortunately this is not a problem for us, and we can identify several possible consequences of climate change for British wildlife that are open to investigation by standard scientific methods. Evidence is always central to good judgement, but this is especially true of climate science, where the public and politicians, not just other researchers, need to be convinced.

Phenology

Phenology, the study of the timing of events in the life-cycles of plants and animals, is perhaps the best-known subject of climate change investigations. Temperate countries with distinct seasons through a calendar year are ideally placed to look for all sorts of variations, from the first snowdrops in spring to leaf-fall in autumn. In a casual way it has long been a British tradition. For decades past, national newspapers have printed letters, every April, from people claiming to have heard the first Cuckoo *Cuculus canorus* of the year. Is this the basis of an ideal data set to study climate change? Maybe not, because migration from distant lands complicates any relationship with what is going on in the UK. Cuckoos could be influenced on their travels by events far distant from our shores. On top of that, the declining abundance of this Machiavellian visitor makes first-date records increasingly unreliable now that there are fewer opportunities to listen out for them. Subjects for phenological study need careful scrutiny to minimise influences other than local climate. In general, though, we might expect a warming environment to affect the timing of life-history traits, particularly of plants and of ectothermic animals that respond sensitively to temperature.

Phenology is therefore a potentially useful tool with which to look for significant, climate-related variations. It's also a field in which the UK can claim an unrivalled track record, largely due to one man and his family living in an east Norfolk village in the 18th century. Robert Marsham (see p. 244), way ahead of his time, started to keep annual records of 27 spring events almost 300 years ago. From 1736 until his death in 1797, Marsham noted the first snowdrops in bloom, the first butterflies on the wing, the leafing times of 13 tree species, breeding activities of frogs and toads, and much else besides. His family continued this tradition right through to 1958. Today's scientific literature, boasting 'long time series' of perhaps 30 years of records, might doff its collective hat to one man's personal efforts that extended over twice that time-span. Marsham's legacy of phenological data was reinvestigated by Tim Sparks and his colleagues at the UK's Centre for Ecology and Hydrology (CEH) and proved remarkably valuable in relation to what is going on now, as described by Sparks & Carey (1995). Others wishing to celebrate Marsham's contribution could do worse than visit the pub he founded for farm labourers, and which bears his name, in the village of Hevingham close to where he spent all his adult life. Robert Marsham's family has given way to some notable successors in the 20th and 21st centuries. Jean Combes has recorded leafing dates for trees in Surrey since 1947, and Mary Manning has made notes in her garden, not far from where Marsham lived, since 1965. These and other dedicated enthusiasts continue to make significant contributions to phenology, providing numbers for crunching by those who delight in such exercises. And many of us do.

Phenology has the great advantage that almost anybody can contribute information with minimal requirements for training. It has become increasingly popular since the creation of the UK Phenology Network (UKPN), with a website (www.naturescalendar.org.uk) hosted by the Woodland Trust where records can be submitted online and where up-to-date information about a wide range of species is readily accessible. This replaced an older recording scheme run by the Meteorological Office, and now has tens of thousands of recorders submitting information. Comparable schemes have sprung up in other

Large White butterflies are among the commonest and most easily identified British insects, and are therefore ideal subjects for phenological study.

countries; in Ireland, for example, Trinity College Dublin's Centre for Biodiversity Research includes a phenology research group. Provided species can be identified correctly – and most of the good indicators are well known, requiring little expertise – it is surely an ideal subject for 'citizen science'.

Almost inevitably, though, there are caveats. A good example of the care needed in collecting phenology data is watching out for the first Common Frog *Rana temporaria* spawn in spring, on the face of it a simple thing to do. Frog spawn is distinctive as gloopy, tapioca-style clumps and is usually laid in shallow water where it is easy to spot. But which pond do you look at? Frogs often breed in many ponds within a given area, and as Malcolm Smith, a revered British herpetologist of the early 20th century, noted: 'the dates of spawning in different ponds in the same part of the country are by no means uniform. Over a distance of a few miles they may differ by as much as three weeks. Differences ... may occur in ponds only 50 yards apart.' These are big intrinsic variations when the aim is to detect trends due to climate that may generate changes of just days per decade. The same issue arises for many other phenology indicators. How to get round this problem? The simplest way is to make sure all records are taken from the same site, in this case pond, on the basis that everything except climate should be more or less constant over time. But it might not be. The pond environment could alter, perhaps with more shade from trees, which would cool the water and delay the frogs. Another apparently trivial but potentially confounding issue concerns the frequency of site visits. I can watch for the first frog spawn in my garden ponds by merely walking out of the back door, but visiting countryside ponds requires more effort and will likely happen less often. Just as with different ponds, the differences in precision between daily and perhaps weekly checks are important when looking for changes that might be, on average, no more than a day or two per year. At the very least, site visit frequencies should aim to be consistent over time to provide the most useful information.

Phenologists need to keep a wary eye out for factors like these that might compromise the interpretation of climate effects. A good plan for an individual observer is to monitor several sites and hope that at least some survive with little change into perpetuity, or at least into the recorder's dotage. Happily, reservations about local variation can be addressed by collecting information from many different people and sites in a locality, and hoping that trends will be detectable by averaging it all out.

Another important variable, this time an interesting rather than a problematic one, is geographical location within the UK. Like many other signs of spring, frog spawning times are inherently climate-related, and have always been earliest in the mild south-west of England and progressively later moving north and east. It's been said that spring moves at a walking pace, judged by the appearance of daffodil blooms, from the south of England to the north of Scotland. A sound, large-scale phenological study therefore requires information from every part of the UK. Given a national database it is possible to tease out not just whether events in a particular area are changing but also how changes map across the country.

These practical issues and statistical methods for interpreting phenological data in the context of climate change have received a lot of critical attention, as described by Hudson & Kealey (2010). Most attempts to interpret phenological changes are based on regression analysis, essentially correlations between the timing of an event (such as first frog spawn) and a climate variable. This variable is usually average temperature in one or more of the months preceding the observation, though it could be something else, such as rainfall. Multiple regression allows the inclusion of more than one variable (maybe temperature and rainfall together) as well as estimating their relative importances. Temperature has dominated analyses of spring phenology in the UK largely because, at least for plants, winter rainfall is deemed unlikely to be a limiting factor. For other organisms, such as amphibians, ignoring rainfall patterns may be unwise.

Regression is one of the simpler statistical tools, but despite this appeal there are some potential complications with it. One is deciding where to begin and end a time series, which can critically influence the strength of any correlation because the method assumes a simple, linear relationship between (say) frog spawn date and year of observation. However, life is rarely linear, and climate change has certainly not been. Looking just at the 1990s shows, for many species, a much stronger correlation and faster rate of change than homing in on the decades immediately before or after that period. A mere glance at graphs of phenological changes is often enough to make it clear that over long periods, running into decades, they have not been uniform. Despite this common knowledge, many analysts have ignored non-linearity on the basis that it complicates matters, and that if there is an overall, statistically significant correlation across several decades (even if the line looks a bit bendy) the point is made. This is fair enough, but by

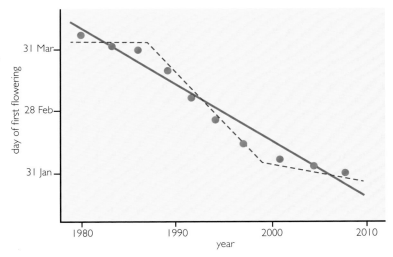

Regression of flowering dates against year, using hypothetical data.

—— simple regression of all data points

– – – more realistic trends over three different time intervals

ignoring the bends we are missing something. If there is an obvious time when the graph changes shape, as in the hypothetical example above, what was happening then? Did climate warming speed up, or slow down? We can look either side of that time to identify two different rates of change, but the decision as to where to draw the line can be difficult, because the transition is rarely a sharp one. With standard statistics, that is pretty much all we can do.

As it happens, there are other ways, as well as regression, to investigate the significance of phenological changes, and the most powerful of these has a wonderfully esoteric origin. Thomas Bayes was an 18th-century minister of the church who found time to 'solve a problem in the doctrine of chances', a startling piece of work that secured him a place as a Fellow of the prestigious Royal Society in 1742. Bayes was, apparently, 'not evangelical in his doctrine' and not a popular preacher. If his sermons were as difficult to follow as the theorem that bears his name, this is hardly surprising. Bayes's contribution to statistics lay fallow among biologists for more than two centuries, but has recently come of age. Statistical methods such as regression, which most ecologists have long been familiar with, give a straightforward yes or no answer; an effect is either significant or not. Bayesian analysis is less constrained and generates a range of probabilities for different possibilities. For example, going back to the issue of regression lines that appear to be non-linear: let's say the increasing earliness of snowdrop flowering seems to speed up at some point around the early 1990s. Was that a real effect? If so, where exactly was the point at which the relationship changed? The options go beyond yes or no, the only ones offered by regression based on a subjective decision about when the change happened. Bayesian statistics give different probabilities across the time period where there looks to be a transition, allowing us to choose which is the most likely flexion point. As mentioned above, phenological changes usually do not 'look' linear, and Bayes's method can be hugely helpful in confirming or dismissing such suspicions. It is also possible to see whether another variable, usually temperature, made a rate-of-change transition at the same time as the flowering event. If so, using Bayesian statistics to point the finger at climate warming as the cause of earlier flowering carries more weight than relying on simple regression alone.

Despite their increasing popularity, Bayesian methods are more difficult to apply than the old-fashioned sort, as there are fewer software packages available for those of us who are mathematically challenged, and they remain relatively rare in the wildlife and climate change literature. Even so, there can be little doubt that Bayesian approaches will become increasingly popular in future. The important point, though, is that, irrespective of the statistical method used, we can now confidently translate what once seemed like casual observations of phenological change into robust trends of national significance.

Species distributions

In addition to phenological change, another prediction of a warming climate is that species currently focused in southern Britain, or present in nearby continental Europe, might extend their distributions further north or colonise the country for the first time. Conversely, northerly, or cold-water marine, as well as montane species might retreat to higher latitudes or altitudes. As with phenology, the UK is well placed to detect range shifts because of its unparalleled biological recording history, extending back to Victorian times and sometimes beyond. Very large data sets for a wide range of species have been collected and assembled for reference on the National Biodiversity Network (NBN) gateway (www.nbn.org.uk), a facility that is available for public access online. Recent surveys can therefore be compared with historical information to assess whether any particular species has increased or decreased its range. Walking in the countryside can become a detective story, always on the lookout for unexpected newcomers or wondering why you now have to climb higher up the hill to find those favourite upland plants.

Swallowtails are rare, beautiful and well-monitored butterflies in the UK, so any change in their distribution should be readily noticed.

However, as with phenology, the study of distribution changes is not always as simple as it sounds. Undoubtedly the biggest problem concerns differences, usually large increases, in recorder effort over time. Widespread participation in local and national recording schemes over recent decades means that we are not comparing like with like. Absence of a species record somewhere during the 19th century, but its discovery there in the 20th, does not necessarily imply a range extension. Perhaps nobody was looking in exactly the right place 100 years ago. At the very least, the situation may be equivocal. The Lesser Silver Water Beetle *Hydrochara caraboides* was well documented in the Cambridgeshire Fens, London Basin and Somerset Levels until the early 20th century, but very rarely from anywhere else. By mid-century it had apparently retreated, due to habitat destruction, to a last stronghold in Somerset. Then, to collective surprise, a breeding colony was discovered in several ponds on the Cheshire plain and in nearby parts of Wales in the 1990s. The area was far from the older records, and allegedly well-worked by entomologists over many years before the discovery. Was this a sudden range extension? Well, perhaps not. Frank Balfour-Browne, a doyen of water beetle enthusiasts, reported back in the 1950s that 'a specimen was found one night in 1903 on a greenhouse roof in the Southport district of south

The distribution of the Lesser Silver Water Beetle is open to interpretation.

Lancs' and another from 'a dry dyke near Little Eaton, Derby'. The beetle is a frequent flier and these places are not so far from Cheshire. Maybe it was always present somewhere in that part of England. In any case, it is excellent news that the fate of this rare insect does not now rest solely in one part of Britain. Seeing one crawl out of a mass of weed, dredged up in a pond net, is always a heart-fluttering moment.

Fortunately, the problems associated with comparing old and new distribution data are not insuperable and have received critical scientific attention. One solution proposed by Telfer *et al.* (2002) is based on comparing records from the same sets of grid squares at (say) two different times, for multiple species within the same taxonomic group (birds, butterflies or whatever). After a statistical transformation that need not concern us, the number of squares where each species occurs is plotted as one time set against another. Thus, in a simplified hypothetical example (see below), Orange Tip butterflies *Anthocharis cardamines* may have been reported in 20 squares in 1950 but 45 in 2010; Brimstones *Gonepteryx rhamni* in 30 squares in 1950 but 68 in 2010; Large Whites *Pieris brassicae* in 40 squares in 1950 but 90 in 2010; while Swallowtails *Papilio machaon* were recorded from 9 squares in 1950 and 36 in 2010. How can we interpret these numbers? The three commoner species lie on a straight line, with recent records for all of them two and a quarter times more numerous than earlier ones, while the Swallowtail, with a fourfold increase in records, is an outlier. This suggests a general increase in records for everything, just meaning more recorders, but a relatively greater increase, indicating

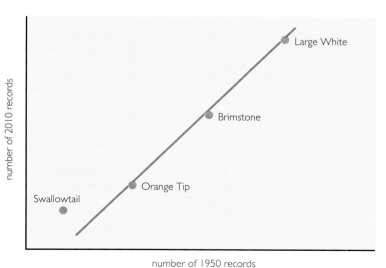

Comparing hypothetical old and recent butterfly records to detect real change.

range expansion, for the rarity. There are some obvious limitations of this method. It requires a large data set of multiple species and generates relative, rather than absolute, inferences of change. It also has weak statistical power, meaning that real changes can easily be missed. There are alternatives, most of which try to compensate for different recorder effort by random re-sampling of the most intensive (normally the later) period data, then making comparisons based on equal numbers of recordings in the early and late periods. Versions of this approach differ in detail, but the method can give a statistically robust indication of range changes, which of course is what we need, as suggested by Hassall & Thompson (2010). Evidently detecting a real range change is more difficult than it seems at first sight.

When we think about recording a plant or animal, it is usually in a positive context. It's a personal high to find an exciting species in a new location, and that is how range expansion gets noticed. Some such discoveries are truly remarkable and reflect the impressive range of identification skills to be found among British naturalists. Stephen Hewitt must have been well pleased to find the first *Platypalpus aliterolamellatus* ever seen in the UK. This small predatory fly (just 2mm long) inhabits river shingle banks and requires a microscope to confirm identity. But there is another side to recording: not finding something that should be there. Negative records are dull, too often not registered at all, but are just as important as positive ones to assess overall range changes. How can we tell when a species has really gone missing, maybe extinct, rather than has just eluded our spotting skills on a particular visit? There are still people looking out for Tasmania's Thylacine even though this carnivorous marsupial has almost certainly been extinct since the 1930s. The internet is loaded with 'reliable' sightings: the Thylacine Awareness Group has several thousand members 'dedicated to the research, recognition and conservation of our most elusive apex predator'.

An alternative to desperate wishful thinking is a scientific method that determines detection probabilities. What are the chances that Great Crested Newts *Triturus cristatus* were present in that super-looking pond, but you failed to find them? This probability can be measured, based on the idea that if you visit a pond several times (let's say four), but only see the newts on one occasion, it looks like your luck will only be in a quarter of the time. From this basic information, survey protocols can be designed to show how many negative visits you need to make to be (say) 95 per cent certain that the newts really aren't there. Negative records are not only disappointing, but are also particularly difficult to make convincing. That doesn't make them any less important, particularly for species that may be retreating in the face of climate change.

Investigating range changes clearly needs careful planning such that statistical tests work effectively and produce credible results. Records from amateur naturalists make useful contributions and are regularly targeted by organised research projects. Many such schemes, often run by non-government organisations (NGOs), are in operation. Buglife, for instance, was at the time of writing keen to obtain records of British oil beetles (*Meloe* species).

It is now possible for individuals to submit records directly to the NBN gateway; links are provided on their website (see p. 27). Irrespective of the science, it's always fun to keep an eye open for animals or plants outside their recognised range. Of course, range changes often have nothing to do with climate. Extra evidence is required to discount other causes, commonest of which is likely to be habitat alteration or destruction associated, usually, with range contractions.

Abundance

Another predicted consequence of climate change is that population sizes of temperature-sensitive species may increase or decrease according to circumstance, whether or not there are changes in overall range. Survival rates might alter as winters become milder, or breeding may become more successful in warmer summers. New habitats, previously lacking appropriate microclimates, might be colonised. Detecting changes in population size, density or extent even at local scales requires considerable effort, albeit more for some species than for others. Counting relatively conspicuous Fragrant Orchids *Gymnadenia conopsea* at a discrete site on the South Downs was repeated annually for many years by Mike Hutchings, alone or with occasional assistance. Unfortunately, assessing changes in the abundance of most species is rarely as straightforward as in the case of orchids. Measuring population sizes of most species, let alone changes in them, with sufficient scientific rigour to convince colleagues is not for the faint-hearted. Long-term investigations, especially at a national level, are inevitably labour-intensive and depend on combinations of professional ecologists, trained volunteers and funding levels that most ecologists can only dream of.

Deciding which species are worthy of attention for assessing population changes, given the likely costs involved, is therefore essential. One rare animal expected to do well in a warming climate is the Sand Lizard *Lacerta agilis*. Britain is at the northerly edge of this reptile's range, and it is confined to warm, unshaded habitats, mostly lowland heaths and coastal sand dunes. In central Europe a much broader variety of habitats is occupied and the critical factor is probably sufficient warmth for egg development. There are clearly possibilities for increases in population density if hatch rates improve, or extension to new habitats if they become suitable for egg survival. There are no data on Sand Lizard population dynamics in the UK, but a long-term study in the Netherlands, where Sand Lizard habitats and ecology are very similar to those here, proved illuminating and was described by Kery *et al.* (2009). A national monitoring scheme with several hundred volunteers, surveying lizards at more than 200 locations across the country, revealed that this attractive animal had indeed increased, over at least a decade, on both heath and dune habitats. Climate warming seems a plausible reason for the lizard doing better, because dunes and heaths in the Netherlands are, as elsewhere in north-west Europe, not increasing in extent or improving in quality; more likely the converse. Why do we know what has happened to Sand Lizards in the Netherlands but not in the UK? Essentially this is because the Dutch government put up the money. In the UK we rely more on wildlife NGOs to do this kind of job, and only the richest can afford the expensive research programmes required.

Sand Lizards might benefit from a warming climate, and systematic monitoring of this attractive reptile in the Netherlands suggests that in many places their populations are increasing.

What's going on?

We need to be cautious when investigating population changes even when the evidence for them is sound. Increased population sizes, and ranges, can happen for reasons unrelated to climate even in a country where most good habitats are deteriorating. The rapid spreads of some introduced species are among the most dramatic examples, but there are others unrelated to exotics. Bitterns *Botaurus stellaris* have a chequered history in the UK, going extinct in the 19th century, recolonising but then declining almost to extinction again in the 20th century, and finally increasing some tenfold since 2000. But we can be sure that climate has had little to do with this bird's recent success. Instead, it has followed an all-too-rare example of extensive and effective habitat improvement. A major increase in reedbeds, entirely due to dedicated conservation efforts at many locations across the country, has done the trick. On the Somerset Levels, Bittern numbers have risen from nothing to the largest population in the UK, all due to large-scale reedbed creation. If anyone was doing the sums, a similar picture of national population expansion would surely be apparent for the Common Reed *Phragmites australis*, harbinger of the Bittern's comeback. There are now more than 5,000 hectares of reedbeds in the UK, more than at any time since the fenland drainage operations of past centuries. Many examples of non-climatic influences on abundance and distribution are less obvious than the Bittern case, and establishing a credible correlation with

BELOW: Common Reed, a plant that has spread dramatically, in concert with birds such as Bitterns, as a result of deliberate habitat management.

Bitterns (inset) are a conservation success story. These birds have increased in the UK because of extensive reedbed creation, not as a result of climate change.

Little Bitterns also use reedbeds but, unlike the native Bittern, increasing visits to the UK of this southern European summer migrant may relate to improving climatic conditions.

climate factors is essential to make that link. Intriguingly, the recent occasional breeding of Little Bitterns *Ixobrychus minutus* in England might well be climate-related. This is a species of southern and central Europe, and a summer migrant, but it also uses reedbeds – so again it could simply be benefiting from habitat improvements.

Communities

Arguably the most profound consequence of climate change where wildlife is concerned will be irreversible disruptions of longstanding ecological communities, but this is a difficult subject to investigate thoroughly. Community ecology is one of the most complex subjects in biology, because ecosystems harbour many disparate components with a concomitantly huge number of interspecific interactions. Scientists from other disciplines who occasionally disparage ecology as a 'soft' science, and there are some, should give community ecology a go. Expertise in the subject requires a daunting combination of species identification and multivariate statistical skills, far removed from most amateur abilities. Experimental studies in community ecology can impart a strong feeling for the bloody-mindedness of nature. Ask anyone who has created artificial ponds to investigate how freshwater organisms interact. 'Identical' replicates of each treatment, essential for subsequent statistical analysis, often make chalk and cheese seem like bedfellows by the end of the study; one thick with algae, another dank and dismal, all due to random events impossible to avert. How, then, can we make sense of communities in the real world without even a nominal control of the possible variables?

At one level, particularly for plants, assessing community structure and change should be possible where discrete patches of habitat can be investigated using standardised quadrats and species counts. Even in this situation, obtaining the counts is one thing but understanding what they mean in a climate context is quite another. Just as with phenology, distribution and abundance, factors other than climate commonly impact on community-level changes. A grim example is eutrophication (enrichment of soil or water with excess nutrients) from external nitrogen input, which has been damaging habitats in the UK for decades and generally leaves a much stronger signal in plant communities than climate change. Many ideas currently doing the rounds about climate effects within communities relate to disconnects in phenology, with some species responding quickly and others more slowly, or not at all, to changing conditions. This kind of study attempts to simplify communities by homing in on two, or a very few, of the component species. A much-vaunted example is a possible discordance between tit breeding times and caterpillar abundance, implying that tit chicks could go hungry and breeding success drastically diminish. More of that in later chapters, but given the difficulty of proving or disproving even such simple community effects, a comprehensive understanding of climatic impacts at the ecosystem level may be a long time coming. Apart from anything else, community changes typically take longer to materialise than anything that happens to a particular species. Interspecific competition, where organisms strive against each other to survive, is often the most important driver in community dynamics, but it usually takes a while for species composition to change sufficiently to be readily detectable.

Monitoring climate change in the UK

Although most government interest in climate change has concentrated on aspects that might threaten human wellbeing, wildlife has not been completely ignored. The Environmental Change Network (ECN), established in 1992, is coordinated by the CEH with a remit to monitor multiple sites all over the UK. Its scope covers any and every cause of environmental alteration, but part of its work involves monitoring three prospective climate change indicator species: one moth and two butterflies. On top of that, a composite biodiversity indicator has been derived based on trends in 21 butterflies, 115 moths and 15 carabid beetles, all considered sensitive to climate change. The CEH programme took on board the results of trying out a wide range of 'UK climate change indicators', which mulled over 34 possibilities between 1999 and 2003. A few of these were wildlife-related but most were not; they included such gripping subjects as the future of domestic property insurance and increases in food poisoning. An interesting conclusion was that warmer summers are unlikely to extend human longevity but will increase the chances of dying in that season.

Then came the UK Climate Impacts Programme, based at Oxford with an acronym incongruously similar to that of a political party hardly renowned for its green credentials. Not much about wildlife here; one of its few studies claiming to be wildlife-related was 'creating sustainable drainage in a new housing scheme', but on the plus side the programme does have a recipe section for cakes and puddings. There is now an Environmental Change Institute at Oxford University, including a website littered with terms we have come to know, but not love: nature-based solutions, natural capital, green infrastructure, ecosystem services, that kind of thing – more than enough to send me out rambling on the coldest winter day.

Nevertheless, valuable information has emerged over the years, including an ever-accelerating publication rate of scientific papers on topics relating to wildlife and climate change. A useful starting point

Common Blue butterfly, a climate change indicator species.

when thinking about this book was the Natural Environment Research Council's (NERC) *Climate Change and Biodiversity Report Card*, published in 2015, which summarises an impressive amount of serious scientific study. Take a look: www.nerc.ac.uk/research/partnerships/ride/lwec/report-cards/biodiversity. While there, also take a look at the frog picture relating to spawning phenology in the composite scheme shown on p. 28 of that report. It's just not British. There has, in addition, been a range of other report cards published by various government agencies over the years, documenting evidence for impacts of climate change on different taxonomic groups and habitats, and in different geographical regions of the UK.

Overview

In this chapter I have outlined several ways in which climate change could affect wildlife in the UK, but I have also emphasised how making any such link can be fraught with difficulties, mainly because other, potentially confounding, factors are usually also at work. In subsequent chapters I delve more intimately into the subject to recount true-life stories, mostly based on serious scientific investigations, concerning how climate change has already impacted the phenology, distribution and abundance of plants, animals and fungi over recent decades. Community aspects, which include contributions from multiple taxonomic groups, are considered in a series of dedicated chapters, including one about our coasts and seas. I have, until the penultimate chapter, avoided published studies (of which there are many) based on computer simulations of what might happen in the future. Such extrapolation is certainly important, not least in the vain hope that politicians might sit up and take notice. Even for this purpose, though, existing evidence puts decision-making on safe ground and should suffice to trigger responsible reactions. Nobel Laureate Nils Bohr's comment that 'prediction is very difficult, especially if it's about the future' should be kept in mind even in this era of computer modelling. Nevertheless, some attempt at prediction must be made, and in the last two chapters of this book I conclude with a discussion of what the future (perhaps) holds and how conservationists might respond. This is not by any means a hopelessly gloomy prospect. At a time when so much of British wildlife is in decline, we are welcoming newcomers that arrive here naturally by flitting unaided across the English Channel, and watching some old favourites extend their range into distant northerly outposts. There is good news as well as bad for future naturalists.

How are plants responding?

chapter two

Unlike mobile organisms, plants cannot as individuals evade the consequences of an environment that alters around them. Like aristocrats in a cultural revolution, they must adapt or die, and in the context of climate change adaptation, this normally means changing the timing of life-cycle events such as flowering or fruiting to accommodate the new situation. This, of course, is what phenology records are all about. At the population level, a longer-term response can involve changes in abundance and/or distribution, for example with a range shift northwards or by successful invasion of previously unavailable habitats.

The UK's green carpet comprises more than 5,000 species, a potpourri of vascular plants, bryophytes (mosses and liverworts) and algae. Astonishingly, only about half of the nearly 3,000 vasculars (flowering plants, conifers, ferns, horsetails and clubmosses) are native. The rest have been introduced, over several thousand years, by humans. Britons' love of parks and gardens, coupled with almost non-existent import regulations, no doubt accounts for much of this unplanned biodiversity bonus. In addition, the UK is home to about 750 species of mosses, nearly 300 liverworts and four hornworts. There is therefore no shortage of candidates when it comes to searching out possible climate change responders.

Fortunately, as well as the army of amateur and professional naturalists contributing to the *Nature's Calendar* website, there is also a bountiful supply of dedicated botanists recording changes in the UK's flora. These folk have contributed over the years to national distribution databases including that of the Botanical Society of Britain and Ireland (BSBI; http://bsbi.org).

OPPOSITE PAGE:
Wildflower meadows like this one, now sadly rare, are a vivid demonstration of the UK's floral heritage.

In this chapter the focus is on particular plant species, groups of plants and some lichens. The responses of other lichens and fungi are investigated in Chapter 5, while the wider issue of changes in community composition, which includes more about algae, is considered in Chapters 6 and 7.

In practical terms, plants have some generic advantages for climate studies, especially for phenology. Immobility is an obvious plus: we can revisit the same area, indeed often the same individuals, year on year and for many species be pretty sure they will still be in place. Not always, of course. Meadows are ploughed up, fields go under housing estates and so on, but, with care and a bit of luck, complication by these unhappy prospects can be avoided. For anyone delighted by the onset of a new spring (and who isn't?), this is a setting in which gardens really come into their own. Those first Snowdrops *Galanthus nivalis*, tantalising glimpses of future warmth, double up as data points on a flowering-time graph. Not as romantic or heart-warming as the sight itself, but such an easy statistic to gather year on year. Further afield, trees are among the best candidates for long-term studies. Individuals are easily recognised over decades, and many species have life-history traits ideal for recording life-cycle events, including leafing times, flowering times, appearance of fruit and leaf-fall. A wide range of smaller plants can be studied in the same way, though not usually on an individual basis. All of which sounds straightforward, but scientists always need some nagging complication to worry about. Reputable naturalists help to sustain this anxiety, and an important qualifier for interpreting a potential data feast was recognised well over a century ago in Richard Jefferies' remarks on field lore in *Wild Life in a Southern County*:

> *Spring dates are quite different according to the locality, and when violets may be found in one district, in another there is hardly a woodbine leaf out. The border line may be traced, and is occasionally so narrow one may cross over it almost at a step.*

This problem is the same as that mentioned in Chapter 1 concerning frog spawn, and can only be overcome by averaging multiple records from many different places.

What about range or population changes? Botanists have been more cautious than entomologists in ascribing range changes, in particular, to climate effects. This is because range adjustments often have other causes, and can be constrained in a variety of ways. In this

Late snow over Snowdrops, one of the plants flowering earlier in response to climate change.

case, the limited dispersal ability of many plants can be a significant disadvantage. Habitat specialists strongly dependent on substrate type may be too distant from any suitable unoccupied habitat to respond sensitively. Dorset Heath *Erica ciliaris* would need to make some impressive hops, at least tens of kilometres, to reach an appropriate substrate where it might extend its range significantly. Competition from species already established might also retard invasions. Pollen, whether carried by wind or insect, is often highly mobile, but this will not of itself extend a range. Seeds are the key to this, and seed mobility varies enormously among plant species. It's hard to imagine rapid range extension of Horse-chestnut trees *Aesculus hippocastanum* without the help of human hands, deliberate or accidental. What happens to unused conkers discarded by children moving on to computer games?

Population size and range are closely linked in most plants, and population increases without simultaneous range expansion are relatively rare. However, the converse is not necessarily true. Population declines can occur from climate-related factors such as temperature or water stress, or from diseases that become more virulent due to changing conditions, without any impact on overall range, at least in the short term. Sick trees, in particular, are an increasingly common sight in the British countryside, and in some cases the changing climate might play a part. This potential threat is discussed more fully in Chapters 5 and 6.

Horse-chestnut 'conkers' are not adapted to unaided long-distance dispersal. This one is germinating on the isle of Rum.

Phenology

Peering into history

In many respects, plants are the stars of phenology. Those selected by the UK Phenology Network are reasonably easy to identify and, of course, never run or fly away, an attractive attribute that several recorders have pointed out. Precision is consequently higher than, for example, reporting the first Swallow *Hirundo rustica* – which might be seen days after another had passed the window unnoticed. For some people a garden is the perfect habitat to work in, with the added bonus that age and infirmity do not necessarily call a halt to enthusiasm. Just looking out of a living-room window, with rain and gales lashing outside, can provide the most comfortable of data points. As of 2016, the Network was collecting information about seasonal events for a total of 71 species, almost half of which were plants (see table opposite).

For all the events registered in the table, it is the first instances that normally go on the record. The list is not comprehensive. Reports of 'bud-burst' before full leafing, first tinting and 'full' tinting of leaves as

CLOCKWISE FROM ABOVE:

Bluebell woods, a quintessentially British delight and a springtime event responding to a changing climate.

Lesser Celandine, one of the *Nature's Calendar* subjects blooming earlier as the years go by.

Bud-burst is another spring event eagerly awaited and recorded in trees, including the Alder.

How are plants responding?

Plant phenology measures in the UK

Group	Species	Spring/summer event	Autumn event
Flowers	Bluebell *Hyacinthoides non-scripta*	Flowering	
	Colt's-foot *Tussilago farfara*	Flowering	
	Cuckooflower *Cardamine pratensis*	Flowering	
	Garlic Mustard *Alliaria petiolata*	Flowering	
	Lesser Celandine *Ranunculus ficaria*	Flowering	
	Oxeye Daisy *Leucanthemum vulgare*	Flowering	
	Snowdrop *Galanthus nivalis*	Flowering	
	Wood Anemone *Anemone nemorosa*	Flowering	
Grasses	Cock's-foot *Dactylis glomerata*	Flowering	
	Lawn	First cut	Last cut
	Meadow Foxtail *Alopecurus pratensis*	Flowering	
	Timothy *Phleum pratense*	Flowering	
	Yorkshire-fog *Holcus lanatus*	Flowering	
Shrubs	Blackthorn *Prunus spinosa*	Flowering	Ripe fruit
	Bramble *Rubus fruticosus*	Flowering	Ripe fruit
	Dog Rose *Rosa canina*	Flowering	Ripe fruit
	Elder *Sambucus nigra*	Leafing, flowering	Leaf-fall, ripe fruit
	Hawthorn *Crataegus monogyna*	Leafing, first flowering	Leaf-fall, ripe fruit
	Hazel *Corylus avellana*	Flowering	Leaf-fall, ripe fruit
	Holly *Ilex aquifolium*		Ripe fruit
	Ivy *Hedera helix*		Flowering, ripe fruit
	Lilac *Syringa vulgaris*	Flowering	
Trees	Alder *Alnus glutinosa*	Bud-burst, leafing	
	Ash *Fraxinus excelsior*	Leafing	Leaf-fall, bare tree
	Beech *Fagus sylvatica*	Leafing	Leaf-fall, ripe fruit, bare tree
	Larch *Larix decidua*	Leafing, flowering	
	Field Maple *Acer campestris*	Leafing, flowering	Leaf-fall, bare tree
	Horse-chestnut *Aesculus hippocastanum*	Leafing, flowering	Ripe fruit, leaf-fall, bare tree
	Pedunculate Oak *Quercus robur*	Leafing, flowering	Ripe fruit, leaf-fall, bare tree
	Sessile Oak *Quercus petraea*	Leafing, flowering	Ripe fruit, leaf-fall, bare tree
	Rowan *Sorbus aucuparia*	Leafing, flowering	Ripe fruit, leaf-fall, bare tree
	Silver Birch *Betula pendula*	Leafing, catkins	Leaf-fall, bare tree
	Sycamore *Acer pseudoplantus*	Leafing	Leaf-fall, bare tree

Source: the Woodland Trust's *Nature's Calendar* website.

they change colour in autumn, and complete as well as first leaf-falls are also recorded wherever possible. The *Nature's Calendar* website provides information on identification and also guidelines for maximising record accuracy. Thus, to assess such features as leaf tinting the advice is to wait until the event is clear in at least three trees and only use mature trees more than 30 years old, though for someone not versed in tree craft this might not always be an easy judgement to make. Data reliability is presumably improving all the time as these guidelines are implemented, but the corollary of this is that old data, essential for investigating trends, will be much less precise.

Some possible complications have been investigated, in the hope, I imagine, that they won't matter much. In a study in North America, first flowering dates were positively correlated with population size of the target species in Massachusetts, where spring is a relatively gradual process, but not in mountainous Colorado, where it isn't. There might be a relationship with human population size as well, to which recorder numbers probably relate. The dates were also dependent on sampling frequency, the number of visits made to each site, which needs to be kept as consistent as possible over the years. Methodology details clearly matter and no doubt will always vary to some extent (volunteer recorders are fickle!), but, with any luck, all these problems will only increase noise in the data and not disguise any real trends. Indeed, it could reasonably be argued that trends detected despite these possible drawbacks are especially likely to be significant.

What, then, has been discovered from the accumulated mass of plant phenology information? There are two fundamental questions. Firstly, have some, any or all of the recorded life-history events changed over time? And secondly, for those that have, is it possible to demonstrate a connection with climate?

Early springs

One of the first indications that the seasonal life of plants was changing in the UK came from Richard Fitter's observations, which extended over 47 years at his home at Chinnor in Oxfordshire and were reported by Fitter & Fitter (2002). This was a family affair, not involving the UK Phenology Network (which was still new at the end of the study), and one of many valuable contributions that Fitter and his family have made to natural history and conservation in Britain. Selecting only the most complete of their astonishing 557 data sets,

How are plants responding?

White Dead-nettles in Richard Fitter's garden flowered almost eight weeks earlier in 1990 than they did in 1954.

the Fitters analysed what had happened to 385 species by comparing flowering times during the 1990s with those between 1954 and 1990. The results were dramatic. First flowering dates averaged across all species had advanced by 4.5 days, and for 60 species (16 per cent) the changes were statistically significant. That might seem like a small proportion, but statistics can be overly conservative at times. Many scientists feel that way when they fail to support a pet hypothesis, but in this case the escape clause is warranted. All these results came from just one place, and sample sizes for each species were mostly much smaller than they would be in any study a professional ecologist would design. But here we are dealing with a very different situation, making up for limited statistical rigour by a combination of extraordinary dedication and personal expertise.

Among all the species recorded by the Fitters, two were particularly remarkable. White Dead-nettle *Lamium album* flowered, on average, an amazing 55 days earlier than in previous decades: around 23 January compared with 18 March previously. This surely now classifies as a winter rather than a spring event. It was closely followed by Ivy-leaved Toadflax *Cymbalaria muralis* with a 35-day advance. But the trends were not all one way, and 10 species (3 per cent of the total) flowered later in the 1990s than they had before. Of these, the introduced Buddleia shrub *Buddleja davidii*, so beloved of butterflies, bloomed on average a full 36 days later in the 1990s than in previous decades.

It's not clear why some plants defied expectations, and this sort of anomaly has cropped up quite frequently in other climate change studies. All too often these examples are set aside without further thought in order to focus on majority responses. At some stage the minority species must be taken more seriously. They may turn out to be particularly interesting, since explaining perverse responses should shed light on how climate responses function at the physiological level. In the Oxfordshire garden, changes in flowering times were more marked in annual than in perennial species, and in insect- rather than wind-pollinated plants, perhaps giving some early clues about ecological variations that might influence future community-level trends.

It is all too easy to jump to what looks like an obvious conclusion, namely that earlier flowering was a response to changing weather patterns. To do so, however, would be a leap of faith. Was there actually a demonstrable connection? To address this possible link, trends for each species in the Fitters' study were regressed against temperatures, obtained from the Central England Temperature (CET) records, for 12 months before the flowering events. The connections were made and the ensuing feelings of satisfaction can readily be imagined. Statistics really can be fun when they do what you want. For earliness in spring-flowering plants, the strongest regressions related to average temperature increases in the month previous to flowering, and there was a four-day advance for each increase of 1°C in that month. For summer bloomers the regressions were weaker overall, but the strongest links were with February temperatures. No explanation was proffered for the delayed flowerers, and no associations for any species were reported for rainfall, presumably because temperature was deemed the most likely variable of importance in a country where winter and spring precipitation are unlikely to be limiting factors.

The next question must be: how typical is a single site, in central-southern England, of what has happened in the country as a whole? Commonality was addressed using the UK Phenology Network data set within a decade of its creation, by which time it had accumulated a huge database with input from sites all across the UK. Computing power had, by then, become sufficient to cope with complex statistical procedures scarcely dreamt of in the 20th century. These developments were exploited by Amano *et al.* (2010) using a hierarchical model (an analysis in which different types of data are interlinked) incorporating almost 400,000 records of first flowering dates from 405 plant species,

Wood Anemones are one of many spring flowers on the UK Phenology Network list that have been appearing earlier as winters have become milder.

and time series of variable lengths but reaching back 250 years altogether. The object of the exercise was to estimate community-level trends while taking account of as many possible complicating factors as the available information allowed. Flowering dates from the multiple species were just the starting point. The analysis included inherent interspecific differences in flowering times (Snowdrops always bloom earlier than Spring Crocuses *Crocus vernus*, for example); the different rates at which each species was flowering earlier over time; and the consistent variation in flowering times, every year, from low to high latitudes across the UK. After all that was taken on board, regression of trend in flowering time against temperature was the crux of the matter. The results confirmed and extended those of previous, location- or species-specific studies: the average 'community-level index' of flowering times within the most recent 25 years (1984–2008) was lower, meaning that flowering was earlier, by between 2.2 and 12.7 days, than in any other 25-year period between 1760 and 1984. The strongest correlation of the index was with February–April mean temperatures, with an advance of flowering times averaged across species of about five days for every increase of 1°C. This heroic effort broadly confirmed at the national level what the Fitters had noticed in their Oxfordshire garden. Richard, who died in 2005, would have been delighted.

The evidence demonstrating recent phenological changes in plants now appears overwhelming. What else is there to do? One obvious question concerns what these changes mean for the life and times of the plants themselves. How, for example, does earlier flowering affect the life history of plants later in the year? Is it generally a good thing, perhaps creating a competitive advantage against species that respond more feebly or not at all? Does it lead to earlier or more abundant seed-set? Or maybe it is neutral, with no particular consequences.

On the island of Guernsey, a 27-year study by Bock *et al.* (2014) investigated changes not only in first flowering times but also in the duration of flowering of 232 plant species, again looking for links with ambient temperatures. Does early onset make flowers last longer? Apparently the opposite is more often true. These results mostly concerned garden species, and first flowering times became earlier and average temperatures increased in much the same way as on the British mainland. In this case, once again the garden environment proved its worth because it made collection of extra information about the duration of flowering possible and appealingly easy: just stepping outside the back door is less daunting than keeping tabs on fields or hedgerows. There was an overall trend for flowering duration to decrease, on average by about ten days per decade, while first flowering became earlier by about five days per decade. There was no clear effect of temperature, in any month, on the duration of flowering, which is probably influenced by many other factors such as wind and rainfall. Whether shorter flowering times are adaptive, by allowing early diversion of resources into somatic growth, or maladaptive, by reducing pollen and seed production, remains an open question. This intriguing and somewhat counterintuitive result certainly merits further study. The potential impact of shorter flowering duration on pollinators was not lost on the authors of the study, one of many unpredictable consequences of climate change on complex ecosystems considered in Chapter 6.

Another interesting question relates to how far the advancement of spring events can go. Is there a limit to how early a flower can unfold or a bud set, and if so, what determines it? With autumns and winters warming more than summers in the UK, will the seasons eventually homogenise? Silver Birch and Downy Birch *Betula pubescens* trees contribute, between them, the most abundant tree pollen in the UK. Production typically gets under way by the end of March and its timing is related to average temperatures in that month. Although pollen release has, like so much else, become earlier in recent decades than in previous ones, it may be that a limit to that trend is approaching, as indicated by the investigation of Newnham *et al.* (2013). Physiological factors might account for such a limit. Many plants require a period of winter cold (vernalisation) to prepare for spring growth, but in the case of birch trees British winters still seem cold enough to avoid this kind of limitation. Indeed, pollen release has been weakly but positively correlated with autumn and winter average temperatures. Day length could be the critical determinant that, in the case of birches, has made the pollen release response non-linear, meaning that the rate of change towards earliness has declined and soon may stop altogether. Many plants certainly do respond to variations in daylight hours, but this explanation remains speculative in the case of birch pollen. These considerations are important because they highlight factors that might eventually become constraints on the increasing earliness of other spring events.

Silver Birch catkins may be approaching the limits of earliness compatible with the physiological demands of overwintering.

Summers and autumns are changing too

So much for spring and early summer – but what about the other seasons? Late summer and autumn events are recorded on the *Nature's Calendar* database, but fewer data have accumulated so conclusions are more tentative. It seems that human psychology is at play here: anticipation of spring as the forerunner of summer is keener than watching out for autumn, a harbinger of cold and damp. Even so, there are few more glorious sights than a mixed woodland on the turn, a panoply of all the hues of gold, brown, yellow, red and green that nature has to offer. An autumn walk is surely one of the great joys of living in Britain and no doubt will remain so, but we may be taking our exercise ever closer to Christmas. For recorders, some late summer and autumn events are hard to judge despite the helpful guidelines provided on the *Nature's Calendar* website. When exactly does leaf-fall begin? Yet another problem beckons after making the observations. Can they be interpreted just in terms of simple climate effects?

Late in the year it is less likely to be only temperature that matters. Rainfall, in particular, can be a key determinant of fruiting time and

Autumn is getting later as spring starts earlier, extending the overall growing season. Leaf-fall, here in the Lake District, is also the subject of systematic recording.

leaf-fall. Summer drought promotes early leaf loss while warmth and rain combined may advance fruiting. At least, that seems to be true of blackberries in my part of the world. Nevertheless, changes are afoot and many naturalists would no doubt agree with Tim Sparks's comment that 'anyone over the age of 30 may be surprised by how late some of the first leaf-fall dates have been in the last decade' (Sparks & Smithers 2009). Unfortunately there are very few long-term time series for autumn events, and most date back only to the start of the UK Phenology Network in 1998. One longer-term comparison is possible, though. The original Royal Meteorological Society Network recorded 'bare tree' dates, the time when pretty much all the leaves have dropped, between 1938 and 1946. By the decade of 1999–2008, bare trees were appearing later by 10–18 days in the cases of Ash, Beech, Horse-chestnut and Pedunculate Oak. Mean August–October temperatures between these decades rose by almost 1°C and the average time shifts for this 'bare tree' event were from late October to mid-November. Leaf-fall for these and other trees such as Field Maple, Hawthorn, Hazel, Sessile Oak, Rowan, Silver Birch and Sycamore is now typically not seen until late October.

Evidently a better understanding of autumn phenology awaits longer time series and will likely need complex multivariate statistical analyses, since aspects other than temperature will need to be factored into explanations. We wouldn't want to rewrite Keats's wonderful poem, but an increasing disconnect between mists and mellow fruitfulness towards the closing of the year is clearly on the cards. One consequence of earlier springs and later autumns has been an increase, by almost a month, in the growing season for crops in the UK between the periods 1961–1990 and 2006–2015. For farmers this has apparently proved a mixed blessing, because although crops have longer in which to grow, pests that attack them may also benefit. This ambiguity might at least inhibit politicians from interpreting climate change as an 'ecosystem service'.

Bramble fruiting time in late summer, seen here in Berwickshire, is climate-related, but in complex ways that include both temperature and rainfall.

New approaches to investigate phenology

With the advent of satellite-based remote sensing, phenology entered the space age. This high-tech approach is based on the analysis of reflected light using wavelengths characteristic of vegetation generally, and of plant chlorophylls in particular. The sensors detect the amounts of leaf cover or of photosynthetic pigment density over particular areas, which in turn reflect vegetation growth. This is an important advance in that it permits large areas of ground to be monitored systematically over time, even though it is cruder than ground-based phenology because it cannot recognise individual species events. Development of vegetation measures such as the normalised difference vegetation index (NDVI) and the enhanced vegetation index (EVI), related to green leaf area and total green biomass, allows the satellite photographs to be quantified for inferring phenological events from satellite data (see figure of Scandinavia, overleaf).

Inevitably, the technology is improving all the time. For example, the MERIS Terrestrial Chlorophyll Index (MTCI) measures leaf extent and chlorophyll density separately, increasing the resolution and detail with which phenological changes can be described. 'Ground-truthing', based on information from the same areas collected by the UKPN, has generally confirmed the overviews of vegetation change across seasons revealed from satellites. Analysis of remote-sensing pictures can do more than simply monitor overall changes in vegetation cover and activity through the seasons; it can also detect differences in phenology between vegetation types. Heathland changes show up differently from those of woodland, for example, and even deciduous and evergreen woodlands generate distinctive profiles.

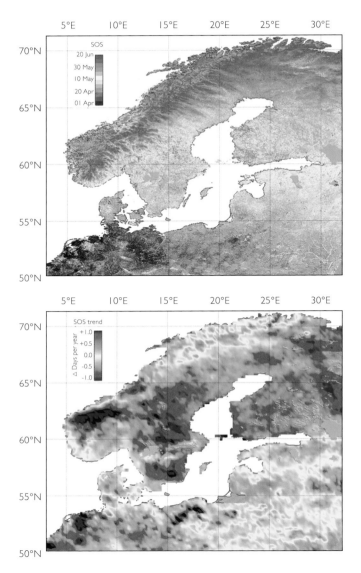

Satellite technology reveals changing phenology at a landscape scale. Arbitrary colours demonstrate differences in the start of the growing season (SOS) across Scandinavia, averaged between 2000 and 2014 (top), and the rate at which the season started earlier during that time (bottom).

Mountainous districts are particularly attractive for satellite-based studies of phenology. Large areas of often inhospitable, even dangerous terrain will always receive less attention from ground-based recorders than the more comfortable lowlands, but mountains are particularly important because arctic-alpine species may be at high risk from increasing temperatures. Much of Scotland is mountainous, and Chapman (2013) investigated satellite data from 2,786 summits there, ranging from 600m to 1,344m above sea level, over the period 2000–2011, to find out what was going on. NDVI and EVI indices were used, together with temperatures in the months preceding each satellite datum point. The vegetation indices were higher in springs and summers of warm years than in cold ones, but the differences did not persist into the autumn. Curiously, the opposite effect was seen in winter. Mountain-tops apparently had less vegetation cover in warm winters than in cold ones; reasons for this remain speculative, and it might, according to the author, be an artefact of the data collection. Evidently the technology may need a bit more refinement. Warmth therefore brought forward vegetation growth in spring but did not significantly affect its rate of decline in autumn, which seems odd and inconsistent with many ground-based observations elsewhere in the UK. The seasonal effects on advancing phenology were greatest at the highest altitudes, and what the consequences of such changes will be for vulnerable arctic-alpine communities must be a matter of concern if such trends continue.

New technical developments in the study of phenology are unlikely to end in space. Drone-based photography shares with satellites the advantage of aerial views but offers the prospect of closer resolution, while also being cheaper and more accessible to amateurs. Drones are already at work on phenology in some countries, particularly the USA, and are increasingly employed in the UK for assessing habitat quality and extent. Let's hope the novel American pastime of shooting drones out of the sky does not take hold on this side of the Atlantic; a woman in Virginia was sufficiently displeased with a drone buzzing over her farm that she took the law very firmly into her own hands: 'It hovered for a second and I blasted it to smithereens.' There are helpful websites pointing out that it is illegal, then telling you how to do it: 'How to shoot down a drone: Don't. But if you're going to ...'

There are also interesting possibilities for using the large arrays of biological records housed at local and national recording centres to investigate phenology. This information is not collected with phenology in mind, but includes a much wider range of taxa than is currently assembled by the UKPN. Computer modelling that takes into account variation of recorder effort over the seasons has shown that, at least for some species, peak (rather than first) flowering times can be estimated accurately from information held in record centres, as described by Chapman *et al.* (2015). No doubt many new insights will follow as an increasing variety of phenological recording methods is implemented and integrated, but for many naturalists the charm of this subject is very personal, keeping an eye out for those first spring flowers or clumps of frog spawn. No technology will ever replace that special pleasure.

Distribution and abundance

Complications

Dandelions come in a wide range of varieties that are difficult to tell apart, confounding attempts to measure changes in their distributions.

Detecting changes in plant distribution and abundance, both potential consequences of climate change, requires a combination of luck, physical effort and good identification skills, happily all characteristics of a good naturalist with botanical bent. However, when changes are small these attributes are barely sufficient; statistical analyses are also required to establish what has really happened from complex data sets. Converting new observations into scientific evidence of climate effects is even more daunting. Systematic comparisons of changes and reasons to believe that climate is responsible for them are needed, and are increasingly being provided, but plants have some features that can make assessment of distribution change particularly problematic. For one thing, species definitions are sometimes extraordinarily difficult. Taxonomic inexactitude is epitomised by the Dandelion *Taraxacum officinale*, a plant with some 200 microspecies that require specialist skills to tell apart. This huge variety of closely related flowers probably arose as a result of their ability to reproduce asexually, such that seeds are usually just clones of the parent. Non-lethal mutations that affect flower structure presumably account for the huge number of dandelion versions we can find in the countryside. This rather goes against the conventional wisdom that asexual reproduction is a bad thing in evolutionary terms because it generally leads to extinction as deleterious mutations accumulate. Maybe dandelions are saved from this fate because sexually reproducing individuals also occur, creating the best of both worlds. From the climate change perspective, though, this is a nightmare species for the study of range changes, because individual microspecies might vary in their responses but their distribution changes go unrecognised because identification is so difficult. Dandelions are an extreme example, but they are not unique with respect to difficulties in deciding what constitutes a single entity whose distribution and abundance can be investigated.

Then there is the issue of non-native species. The UK has about 1,400 vascular plants considered truly native and a further 150 so-called archaeophytes that arrived, presumably by human agency, more than 500 years ago. These can be considered honorary natives because their exact mechanism of arrival in this country, some as far back as several thousand years ago, is not generally known. They

How are plants responding?

include real beauties, many of which were once widespread in hay meadows and cornfields but which all too often have declined rapidly following the advent of agricultural intensification. Among the UK's glorious array of archaeophytes are the Cornflower *Centaurea cyanus*, Corncockle *Agrostemma githago*, Chicory *Cichorium intybus* and White Dead-nettle. Because all of these species have been in the UK for a long time, recent changes in distribution or abundance might reasonably be related to climate. However, there are a further 1,100–1,200 vascular plants established in the wild that are classed as neophytes – relatively new arrivals as a result of accidental or deliberate introductions over the past 500 years.

For the neophytes, the study of range changes can be complicated because, at least for the most recent ones, there may not yet have been time for dispersal to all suitable habitats, irrespective of climate effects. Caution is obviously due here. And there is another problem consequent upon such a large number of alien arrivals. Will it ever be possible to identify species that have established themselves in the UK by natural invasion, exploiting a warming climate, rather than as a result of ongoing human imports? What should we make of the recent discovery of a Sawfly Orchid *Ophrys tenthredinifera* in Dorset? Did this Mediterranean species arrive under its own steam, or was it helped? We'll probably never know, although some clever genetics

LEFT: Cornflowers are well-loved members of wildflower meadows that arrived in the UK sometime in the distant past, almost certainly by human agency rather than as a result of ancient climate change.

RIGHT: Chicory is another glorious archaeophyte attractive to pollinators that has been in the UK long enough to have achieved an equilibrium distribution. Any recent changes could therefore be attributed to climate, but in this case are much more likely to result from habitat deterioration.

LEFT: The Sawfly Orchid is new to the UK, but probably arrived by human agency rather than by a climate-induced range extension.

RIGHT: Sea Daffodils (pictured here at Campania in Italy) appeared recently on the south coast of England and could be natural colonists, though their provenance remains uncertain.

comparing the British plant with putative populations of origin might shed some light on the issue. Arguably more likely contenders for recent, natural arrivals are the Tongue Orchid *Serapias lingua* and the Sea Daffodil *Pancratium maritimum*, both of which are Mediterranean and/or Iberian natives and both of which have appeared at coastal sites in England recently, perhaps via seeds carried in the Gulf Stream. But again it is unlikely, failing genetic work, that we will ever know their true histories. In contrast to some other groups such as grasshoppers, crickets and dragonflies, therefore, there is little mileage in trying to determine the role of climate change in the appearance of new plant species in the UK. Nor is this problem confined to possible recent colonists. There is still much moving about by humans of plants long established in this country. We might choose to blame ancient Romans, modern gardeners and everyone in between for the relatively chaotic state of our floral distributions – but what matters is whether, above this background noise, unequivocal signals of climate change on plant distribution can somehow be extracted.

And there's more. Plants are much more prone to hybridisation than animals, with over 900 hybrid varieties recognised in the UK. Of these, about 300 have at least one non-native parent, including for example the results of crosses between native Bluebells and their Spanish cousins *Hyacinthoides hispanica*, as if invasion by the alien was not annoying enough. Hybridisation may be promoted by both the arrival of new species and by climate change bringing together plants previously isolated by climatic or edaphic (soil-type) factors. Here is another complication, requiring in many cases a high level of identification skill, in the study of climate change impacts on British flora.

What, then, are the signs to look out for in changing distributions or abundance that might be attributable to climate change? A general increase in abundance despite rampant habitat destruction and deterioration is one pointer, providing it is clear that human translocation has not been significant and that, if the plant is a neophyte, the change is very recent after a period of stability. Directional shifts in latitude or altitude, as opposed to occasional new appearances randomly distributed across the country, are another useful guide. Yet another indication could be range infilling, where a species becomes more widespread within its historical distribution and perhaps colonises habitats previously unavailable to it. Despite the difficulties listed above, there is a story to tell.

Losers

A warming world is likely to be good news for many, but by no means all, of the UK's diverse flora. Mountainous regions of northern England, Wales and Scotland are home to subarctic alpine plants that are well adapted to the rigorous conditions of high altitudes and rocky substrates. Some are very beautiful and many are rare on account of their habitat specialisation. Blue Heath *Phyllodoce caerulea*, Alpine Gentian *Gentiana nivalis*, Alpine Bearberry *Arctostaphylos alpina*, Alpine Catchfly

Blue Heath is a rare plant of the Scottish mountains, one of several high-altitude species at risk from climate warming.

Snowdon Lily is another upland gem, in this case confined to its Welsh mountain namesake and also at risk from climate change.

Lychnis alpina and Snowdon Lily *Gagea serotina* are among the possible delights of a highland hike. Increasing temperatures are potentially bad news for these species, which might suffer increased competition from more generalist plants if these start to move uphill, or from desiccation if their world becomes warmer and drier.

One measure of climatic influences in mountainous regions is the timing and extent of summer snow-melt. Relative to the previous 25 years, there were significantly fewer snow patches that survived the summer un-melted in the Cairngorms during the period from 1996 to 2005. In some years no patches at all persisted to the following winter. This extreme environment is home to plants that survive on the margins of snow-beds, including crags and runnels that rely on snow-melt to retain dampness. Data from the Environmental Change Network revealed that temperature changes between 1993 and 2007 were almost twice as fast at high altitudes as in lowlands, a trend that sits nicely with phenological observations but which looks ominous for the future of cold-adapted organisms. In such remote and inhospitable terrain it is inevitably difficult to obtain good enough data on plant distributions to detect long-term trends, but some efforts have been made to find out if the gloomy expectations are being realised.

Montane habitats in Scotland, defined as those above the tree line, generally commence at between 200m and 800m elevation, depending on precise location. At Ben Lawers National Nature Reserve in the southern Scottish highlands, a series of plots was

How are plants responding?

assessed for botanical composition in 2013 and compared with information from plots in the same region in 1950. This was a rare and valuable opportunity to identify changes at a high-altitude site, courtesy of Ross (2015). Seventeen species had declined between the survey periods, ten of which have primarily Arctic or montane distributions. Those showing substantial losses included the liverwort White Frostwort *Gymnomitrion concinnatum*, Dusky Fork-moss *Dicranum fuscescens*, Rock Sedge *Carex saxatilis* and Red-stemmed Feather-moss *Pleurozium schreberi*. Other Arctic/montane plants apparently in trouble, albeit without statistical substantiation, were Marsh Forklet-moss *Dichodontium palustre* and Twiggy Spear-moss *Warnstorfia sarmentosa*. By contrast, a few species were doing rather well. Grasses and some sedges were the main winners, with several increasing substantially in montane habitats including Mat-grass *Nardus stricta*, Deergrass *Trichophorum germanicum* and Common Yellow-sedge *Carex demissa*. Overall, communities had become less distinct over time and there was a homogenisation of species compositions in the higher habitats that could not readily be explained by factors other than climate, such as increased nitrogen deposition or changes in grazing regimes.

A wider-ranging study by Britton *et al.* (2009) across several hundred plots at multiple locations in mainland Scotland, Mull, Orkney and Shetland came to a similar conclusion. Between initial surveys over the period 1963–1987 and resurveys in 2004–2006, 19 out of 61 species declined in their extent of ground cover. These 'losers' were

LEFT: Mat-grass is one of several grasses in upland regions that are benefiting from climate change, extending their distribution and potentially invading the territories of endangered arctic-alpine specialists.

RIGHT: Common Yellow-sedge is faring increasingly well in the Scottish uplands as the climate regime alters in its favour.

The beautiful Trailing Azalea has declined greatly across many sites in Scotland as the climate has worked against it.

again mostly plants with arctic-alpine or montane distributions in the UK, including Stiff Sedge *Carex bigelowii*, Dusky Fork-moss (again), Trailing Azalea *Loiseleuria procumbens*, the lichen *Ochrolechia frigida* and Snow-bed Willow (Dwarf Willow) *Salix herbacea*. Ten 'winners' were mostly temperate species not restricted to high altitudes, including Ling Heather *Calluna vulgaris* but also a couple of unexpected arctic-alpines or montanes, Fir Clubmoss *Huperzia selago* and Whiteworm Lichen *Thamnolia vermicularis*. With a few exceptions, therefore, the pattern of change was largely consistent with the expected consequences of climate warming, but in this more wide-ranging study other factors, especially elevated nitrogen deposition at high altitudes due to the large amounts of rainfall there, may have been significant contributory factors.

There could be a twist in this tale. After 2007, and therefore subsequent to the data gathered by Britton *et al.*, the extent of summer snow-melt at high altitudes in Scotland decreased dramatically, seemingly reversing trends of the previous decades. In 2015 many snow patches persisted and were large enough to generate spectacular networks of ice tunnels and bridges. Ironically, this too could be a consequence of climate change. Increasing winter precipitation in the mountains is likely to be reflected as increased snowfall, as has occurred in parts of Antarctica, and this reversal might assist arctic-alpine plants – as well as human ski enthusiasts – to persist for longer

How are plants responding?

Rare Lady Orchids (left) are improving their lot against the odds of general habitat deterioration, aided by more favourable climatic conditions, whereas Man Orchids (right), like many British wild flowers, are declining.

to climate change, such as quarrying, house building and road construction, might benefit early-successional-stage plants. Equivocal responses by plants in this category include that of Prickly Lettuce, which particularly favours habitats such as roadsides and waste ground. However, this member of the daisy family has been in the UK since before 1632 (its first record) and only made its dramatic expansion very recently, arguably supporting the idea of recent climatic benefits. But maybe the recent distribution infilling by some other species, such as Rough Hawk's-beard *Crepis biennis*, has a more complex cause which might include climate but also its predilection for roadside and railway embankments as well as various wastelands. It is hard to be sure, and a complete list of possible climate beneficiaries would be both tedious and of variable provenance. There are certainly other contenders if large population increases or distribution infilling are accepted as key criteria. They include Southern Marsh-orchids *Dactylorhiza praetermissa*, which have increased hugely on damp, marshy ground in England and Wales; Green-flowered Helleborines *Epipactis phyllanthes*, which have spread from relatively few sites mostly in southern England to many more, especially further north; and Spotted Medick *Medicago arabica*, which has changed from being widespread but local to very much commoner across much of southern and eastern England.

Spotted Medick is one example of several plants that have increased across much of central and southern England by range infilling, probably facilitated by climate change.

Many complex interactions of ecology and life history determine the distribution of any particular species. Time and again, research on plant distribution changes in the UK has emphasised the importance of habitat, all too often habitat deterioration, as the main determinant. A major prediction of climate change is that, habitat permitting, thermophilic (warmth-loving) plants will tend to extend their range northwards as the country warms up. Has that actually happened?

A comprehensive investigation by Groom (2013) of all British vascular plants excepting the rarest, comparing the periods 1978–1994 with 1995–2011, failed to show any trend in that direction specific to thermophiles. On the other hand, northward shifts were occurring, on average, for all the plant species with expanding ranges, but not for those that were declining. Determining distribution shifts for species already common and widespread in the UK is often not straightforward. Many British plants have distributions extending across the whole country, and northerly range shifts for those species obviously cannot be based on changes in distribution boundaries. One method adopted in Groom's study was to look for changes in the 'centre of mass', defined as those areas holding the largest density of records, in the two time periods. Thermophilic plants were faring better than others in terms of increased overall range, but there was no tendency for their centre of mass to move north faster than the overall average for all thriving species.

However, the key fact is that most plant distributions have varied rather little in the UK in recent decades. In another study, by

Doxford & Freckleton (2012) and based on 1,781 widespread species with comparable information available from 1930–1960 and 1987–1999, the median range change was less than 0.2 per cent. Even so, it was interesting to note that rainfall and temperature effects contributed to about 45 per cent of these small expansions or contractions.

Rare species have often changed more dramatically than common ones, as highlighted by examples given earlier. The need for sophisticated statistical analyses to show any effect at all in most cases highlights how little has actually happened to the distributions of most species of vascular plants in the UK.

Despite uncertainties about recording consistency, and the presence of complicating environmental factors, there is evidence that at least a few bryophytes may be benefiting from climate change, as revealed by Blockeel *et al.* (2014). Fingered Cowlwort *Colura calyptrifolia* has increased its range and attained higher altitudes, particularly in Wales and Cornwall, since 1990; similarly, Minute Pouncewort *Cololejeunea minutissima* has spread extensively in southern and central England; Irish Daltonia *Daltonia splachnoides* has increased dramatically in Ireland as well as making new appearances in Wales and Scotland; and Balding Pincushion *Ulota calvescens* has arrived in Wales and north-west England, having previously been confined mostly to western Ireland and Scotland. These are primarily western, 'oceanic' species, likely to do well in warmer and wetter conditions, so climate change is a credible explanation for their recent shifts in range and abundance. Credible does not mean definite, but sometimes, happily, we can set rigorous analysis aside and return to the joys of simple observation. Perceptive naturalists occasionally notice fascinating developments without any attempt at scientific confirmation. Bryophyte namesake Brian Moss recounted that:

> As to anecdotal experiences, the most prominent one, over the last twenty years in the north-west, has been a rise in the moss coverage on lawns and tarmac drives, which I presume has been due to warmer, wetter winters. It seems to have spawned a rise in small local businesses, like 'the Lawn Ranger', concerned with lawn reparation and of business devoted to the clearing of moss. My cross country skis have also been gathering cobwebs in the garage for a very long time.

The same applies to my childhood sledges, indicating not only warming winters but also a less energetic lifestyle than Brian's.

Mechanisms of change

How are climate change responses achieved?

What actually happens to plants as they experience changes in their comfort zone? Initially, we can recognise a distinction between individual and population-level responses. A particular plant may have enough physiological flexibility to, for example, flower earlier or later according to weather in the previous month. By doing so, key life-cycle events can remain within the same climate envelope. Alternatively, or perhaps as well, individuals within a population may vary in their ability to make such a response. This situation could manifest itself if, say, among a cluster of Bluebells some individuals regularly flowered earlier than the rest in a warm spring. At the population level, selection might start to favour early-flowering variants, but there may come a point at which phenological responses are not enough. Individuals in what has become a hostile environment will die, but at the range margins the population may be able to move into more favourable conditions. This will rarely be an option for arctic-alpine plants living under subarctic conditions in the high mountains with nowhere colder to go in the UK, and even for thermophiles there will often be constraints to such a response, owing to limited availability of other key factors such as appropriate soil type.

The responses to a warming climate are therefore complex, because they can include both phenological and distribution changes, or a combination of the two. How has this worked out in practice? In a comparative study by Amano *et al.* (2014) of 293 native plants, 45 per cent showed an advance in first flowering dates between 1930 and 2009, while of 284 species in the same study, nearly 80 per cent had experienced a northward range shift. Perennials were less able than annuals to undergo phenological change, just as the Fitters found in their garden, but perennials were more likely to respond with range shifts. The shorter generation times of annual plants may be a critical factor explaining this difference in climate response, because annuals have only a single season in which to reproduce successfully. Flexible phenological reactions are especially important for success with this lifestyle. Trees, which are arguably the most extreme perennials, however, have provided many examples of substantive phenological change – so any generalisations about lifespan and climate response must be treated with caution. Given the opportunity, some trees can expand their ranges remarkably quickly. Pollen analysis shows that pioneers such as Alder, Silver Birch and Scots Pine moved northwards at up to 500m per year, overwhelming the tundra vegetation as ice caps melted away during the postglacial warming phase in Europe.

Evidently the ranges of most lowland plants in the UK are not changing very much, and the question arises as to how many of them are constrained by climatic rather than other factors, such as soil type or life-history traits. Few trees in the UK, excepting Small-leaved Lime, would be on a climate-sensitive list today.

Of particular interest is whether phenological and distributional changes are linked in some way. Have the marked phenological responses, so obvious in many plants since the 1980s, translated into range shifts or changes in abundance? Further analysis of the Fitters' records failed to find a link between first flowering dates of a species and recent range changes in the UK, so early flowering per se was not a good indicator of expansion prospects. Snowdrops are no better placed to improve their lot than daffodils. However, Hulme (2011) showed that the extent to which first flowering time responded to climate change, as opposed to being intrinsically early, was linked to range changes, mostly increases, between 1970 and 2000. Native species whose phenology did not track climate change tended to decline, whereas neophytes in general responded more strongly than native plants with respect to both phenological and range changes. However, interpreting neophyte responses depends very much on how recently they arrived in the UK. Some of those present for several centuries, such as the Sycamore, have remained more or less static in distribution in recent decades, whereas the herbaceous perennial Dusty Miller (Snow-in-Summer) *Cerastium tomentosum*, first recorded in the wild in 1915, was still increasing dramatically at the end of the 20th century. Evidently at least some recent additions to our flora have not yet reached an equilibrium distribution in the UK, and increases in such neophytes cannot be simply attributed to climate change.

Overall, no systematic differences in ability to increase geographical range between species with wind- or animal-vectored pollen have been seen, suggesting that these dramatically different life-history traits are unlikely to be key determinants of distributional responses to climate change. Time to remember, though, that we are still in an early phase of global warming and drawing firm conclusions about events spanning just two or three decades would be premature. Many tree species have individual lifespans an order of magnitude, or more, greater than this. There are sure to be unpredictable, species-dependent lags in distributional responses – meaning that we have not yet seen the ultimate consequences of climate changes that have already happened. A network of interwoven ecological niches and life-history traits will ultimately determine whether, how and to what degree any particular species of plant will fare in a warming world.

Moving beyond correlation

A not unreasonable criticism of scientific claims about the impact of climate change on British wildlife is the inescapable fact that conclusions are based largely on correlations. Regression and correlation, fundamentally similar methods but with slightly different assumptions, have remained the primary statistical tools available for relating phenology and distribution to climate factors. Unfortunately, both techniques suffer from the inability to prove cause and effect. However strong and statistically significant a correlation turns out to be, the relationship may be 'accidental', or at least indirect. For example, a change in rainfall might cause farmers to grow different crops, which in turn might affect the distribution of a wild plant. Arguably this is still fundamentally a climate effect, but the problem could be worse: climate and distribution might both be influenced independently by something else entirely, an unknown and unmeasured variable hard to imagine in this context, but which can certainly complicate correlation analyses in some situations. The best that can be done with climate is to eliminate other possible causes of change that we happen to know about, on the basis that they should exhibit weaker or non-existent correlations with variations in phenology or distribution. In any case, despite this methodological limitation, climate scientists are in good company. Cosmologists are similarly constrained to observational science, but have successfully convinced us that the sun is about 150 million kilometres distant from the earth without the need for a celestial laboratory. Nevertheless, experimental approaches to complement statistical methods that cannot unequivocally demonstrate cause and effect are a desirable, albeit difficult, goal for future work. Proposing experiments to control climate at a landscape scale is hardly a research grant winner, but there are a few other options that take us at least a little way beyond simple observation.

Genetic studies can provide guidance for identifying potential climate responders, by assessing indirectly how far seeds are typically

Rowan berries are attractive to birds and the seeds within them can travel great distances via avian digestive systems. Genetic studies have confirmed that Rowans have high potential for rapid dispersal.

spread by their various vectors, usually wind or bird digestive systems. One method for doing this is based on molecular markers that show how different one population is from another. Plant chloroplasts contain small quantities of DNA containing genes that can vary among different populations. This DNA is an ideal genetic indicator because it is perpetuated in seeds but not in pollen. Large differences among the chloroplast genes in different localities indicate little gene flow (that is, little seed movement) between populations, and vice versa. This removes the necessity to assess seed movement by direct methods, which can be inordinately difficult. How far will the wind blow, a bird fly or a squirrel scamper? Plants can be ranked on the basis of this genetic information, those with high gene flow being, arguably, the most likely to experience rapid range changes when circumstances permit. Rowan and Alder fare better than Common Beech, which in turn spreads its seeds more widely than Pedunculate Oak or, at the bottom of the scale among those tested, Sessile Oak. This kind of sideways look at biological features that might be important for climate change responses has the advantage of generating predictions about colonisation rates that can be tested in the field.

There is also a place for laboratory studies, more of which are discussed in Chapter 8. Experimentation is not entirely out of the

Sessile Oak acorns seldom travel far from the tree of origin, and genetic analysis has shown that this tree has limited dispersal ability compared with several other species.

question in the study of climate change. A fundamental aspect of phenological change comes down to the seemingly irrepressible nature versus nurture debate. Can we find out to what extent variations in flowering times represent flexible individual responses (phenotypic plasticity, a version of 'nurture') or genetic differences among and within populations (the essence of 'nature')? Obviously, if the timing of flowering, bud-burst or leaf-fall of individual plants varies in different years, the mechanism must be 'nurture'. But there may also be genetic components that could constrain flexible responses and thus impose limits on how far individuals can adapt to climate change.

This is the kind of problem a laboratory scientist can get to grips with. Sessile Oaks in Pyrennean valleys survive over a wide range of altitudes, from 130m to 1,600m above sea level. Seed germination and seedling bud-burst are progressively later as altitude increases, and this gives hope that the trees might adapt to a warming climate with the earlier phenology simply moving uphill. This, however, very much depends on the genetic basis and thus the flexibility of the germination and bud-burst traits. These were tested in so-called 'common garden' experiments by Alberto *et al.* (2011) in which acorns from various altitudes were grown under identical conditions. To all intents and purposes the young trees behaved like their parents, retaining their phenology of origin and confirming strong genetic differentiation up the mountains. That sounds like bad news, a genetic determinism only to be overcome by spreading the better-adapted trees uphill. However, there was also large variation within each area, the last tree bursting its buds at the low altitude coinciding with the first one at high altitude. So perhaps it will only be necessary for early-budding variants to spread within the same area as their parents, which could presumably happen more quickly than longer-distance movements up a valley. Selection based on vulnerability to frost damage probably established the cline of phenology apparent in the oak trees today, but it remains to be seen whether the adaptive traits can spread as fast as warming springs. More experimental studies of this kind, not just with plants, should provide valuable insights about likely future responses as the world heats up.

Overview

Climate change has, so far, been a mixed blessing for British flora. No species have become extinct or look likely to do so in the near future. Most have made only minor responses at the distribution level and a few have improved their lot considerably in the face of widespread habitat deterioration. It seems that we can enjoy the pleasures of early spring inflorescences without worrying too much about long-term ecological consequences, at least in the lowlands. But this is a risky assessment and might prove too optimistic as the changes continue. There is an inherent bias in favour of recording the colonisation of new localities, which is relatively easy and enjoyable, as opposed to documenting extinction, which is much harder. Who can say they have found the last individual of any organism? This caveat is particularly important, because the arctic-alpine plants most at risk are among the most difficult to survey comprehensively. Our high-altitude plants should temper any tendency to wax optimistic about climate change's effects on our floral heritage. Aside from this, though, well-known wildlife writer Richard Mabey's intuitive view seems to hold up:

> *I am by nature a looker on the bright side, fascinated by adaptation and natural resilience. So I regard the impact of climate change on UK wildlife (an important qualification) as largely positive. Subjective from here on in: dramatic increases in the population of all spring and summer flowers – especially more southern species like cowslip and, where habitat is not a factor, almost all orchids.*

Let's hope we don't have to revise this opinion in future.

Invertebrate tales | chapter three

Invertebrates, like plants, have responded to climate change in the UK by altering both phenology and distribution. As ectotherms, invertebrates are critically dependent on external heat sources to sustain activity, and therefore many are potential benefactors of a warming climate. Many, but not all. A good number of them occur in the colder parts of the UK and, like arctic-alpine plants, may be at risk if their world warms significantly.

Excluding marine varieties, more than 25,000 species of invertebrates crawl, scuttle, swim or fly somewhere in the British Isles. These animals exhibit an almost tenfold higher diversity than that of the vascular plants, and more than 90 per cent of our invertebrates are insects. This relative high abundance is manifest in most of the world, prompting prominent ecologist Robert May's comment that 'to a good approximation, all species are insects' – the kind of statement only a mathematically inclined ecologist could get away with.

Although invertebrates have changed both phenology and distribution in response to climate change, the balance between these two aspects is different from that shown by plants. Timing of early spring appearances has been more commonly reported than distribution changes in the plant world, whereas the opposite is true of invertebrates. This may, however, be at least partly an artefact of differential observability. Many plants are easy to see, and noting first flowers or leaves is very straightforward. Slugs, snails, worms, spiders, centipedes and woodlice escape our attention most of the time even though the UK has collectively around 1,500 species of these animals. Who is looking out for the first millipede of the year? The same is true of most insects, the majority of which are hard to find at any time. Nevertheless, there are exceptions, and a few insect groups have

OPPOSITE PAGE:
The Silver-studded Blue, a rare butterfly emerging earlier than it did in the past in its heathland habitats.

Climate Change and British Wildlife

The Brimstone, one of the first signs of spring, is arriving earlier every year.

species that attract public attention sufficiently often as to provide a phenological record. These are invariably conspicuous flyers, especially butterflies. Who can remain unmoved by the first Brimstone fluttering around a woodland glade on a sunny March – or perhaps February – morning, one of spring's earliest harbingers?

Much of insect phenology has been about the first appearances of butterflies that have overwintered as adults and are therefore just emerging from hibernation. Another particularly visible group, though, comprises the Odonata, dragonflies and damselflies, all of which overwinter as aquatic larvae ('nymphs') in ponds, streams and rivers. Immediately prior to metamorphosis, the larvae crawl out of the water and typically up the stem of an emergent water plant, and within a few hours a fully functional adult escapes from the old nymph skin. The 40 or so British species, including migrants, each have distinctive flying periods from spring through to autumn. The appearance of the first adults patrolling the countryside is another notable event that can respond to climate, usually later in the year than butterfly emergence.

The greater mobility of most invertebrates, compared with plants, has led to some dramatic changes in distribution, especially of those species capable of flight. As with phenology, insect records dominate our measurements of invertebrate distribution changes – but their hegemony is not exclusive. Some spiders, with silk strands caught on the wind, travel vast distances without the benefit of wings. Even among the insects there are some unlikely long-distance travellers, including species that, on the face of it, look like cumbersome stay-at-homes. Great Silver Water Beetles *Hydrophilus piceus* are one of Britain's biggest and bulkiest insects, spending most of their lives creeping around in dense beds of waterweeds. As swimmers they look feeble compared with the better-known diving beetles such as *Dytiscus* species, but occasionally Great Silver Water Beetles move on, and in doing so demonstrate remarkable stamina as marathon fliers. Not long ago one turned up on a North Sea oil rig more than 100km offshore – proof that movement of mobile invertebrates can include invasion from the near continent as well as travel within the UK.

General phenology

Invertebrates for which the UK Phenology Network assembles records are listed on the *Nature's Calendar* website (see overleaf). These are vastly outnumbered by species we know nothing about. The invertebrate story highlights the simple truth that for most British species of all animal groups, we have no idea whether their behaviour is being modified by climate change. About the nation's worms, leeches and gastropods (some 220 species), spiders (650 species), centipedes and millipedes (over 100 species) and crustaceans (more than 450 species), there is just nothing to say about changed phenology. There could

Like many other butterflies, Orange Tips are advancing their first appearance dates as the climate warms.

Holly Blues are mid-spring fliers, but they are also on the wing sooner than they were in the recent past.

be some interesting goings-on among these ecologically important beasts. Gardeners and gastronomes alike might want to know, for very different reasons, whether Garden Snails *Cornu aspersum* are benefiting from milder and wetter winters. Are the snails active over a greater part of the year now than was true 30 years ago, are they laying more eggs, and have population dynamics altered to their advantage? These are the kind of things it would be good to discover, but for which the lack of early records probably precludes anyone from ever finding out, although genetic studies could possibly reveal any signs of recent population expansion.

All the species routinely recorded for the UK Phenology Network are flying insects (though ladybirds are not normally seen on the wing) and all are common. Every one occurs regularly in my garden and no doubt in many other people's gardens too, favouring a high response rate from public contributors. Virtually all are distinctive and easy to identify, though the Small White *Pieris rapae* and Green-veined White *P. napi* butterflies require fairly close encounters to distinguish safely. In all but one case it is first appearance after the winter that goes on the record, and in terms of batting order these times are predictable among the species. Brimstones and Orange Tips are early, while Green-veined Whites and Holly Blues *Celastrina argiolus* are typically a month later, so their emergences may be influenced by different climatic events.

Some butterflies overwinter as adults and others as earlier, presumably still developing, life stages. These life-cycle differences might also lead to dependences on distinctive climate features. Red Admirals *Vanessa atalanta* now regularly survive the mild British winters, and though most still arrive as migrants from Europe in late spring or early summer, sightings immediately after hibernation could confuse interpretation of the main trend. Nevertheless, first observations are usually of new arrivals, and the relatively late timing of this glamorous butterfly may be influenced by factors different from those impacting on residents.

The species regularly recorded are the tip of an iceberg, and information about phenology has accrued for many more. Just as with plants, there are useful historical data sets against which recent phenology can be compared. Marsham only noted one insect in his Norfolk rambles, the Brimstone, but newer sources include volunteer recorder input to the UK Butterfly Monitoring Scheme (UKBMS) set up by the CEH and now run in conjunction with Butterfly Conservation. This scheme monitors both phenology and distribution of all British butterflies, with data going back as far as 1976. It therefore covers the critical period of recent climate warming rather well.

Invertebrate phenology measures in the UK

Group	Species	Event
Beetles	Seven-spot Ladybird *Coccinella 7-punctata*	Emergence from hibernation
Bees and wasps	Red-tailed Bumblebee *Bombus lapidarius*	Emergence of queen from hibernation
	Common Wasp *Vespula vulgaris*	Emergence of queen from hibernation
Butterflies	Brimstone *Gonepteryx rhamni*	Emergence from hibernation as adult
	Comma *Nymphalis c-album*	Emergence from hibernation as adult
	Green-veined White *Pieris napi*	Emergence from hibernation as pupa
	Holly Blue *Celastrina argiolus*	Emergence from hibernation as pupa
	Orange Tip *Anthocaris cardamines*	Emergence from hibernation as pupa
	Peacock *Nymphalis io*	Emergence from hibernation as adult
	Small Tortoiseshell *Nymphalis urticae*	Emergence from hibernation as adult
	Small White *Pieris rapae*	Emergence from hibernation as pupa
	Speckled Wood *Pararge aegeria*	Emergence from hibernation as pupa
	Red Admiral *Vanessa atalanta*	First appearance (migrant from Europe)

Source: the Woodland Trust's *Nature's Calendar* website.

General distribution and abundance

Astonishing extensions of invertebrate distributions and increases in abundance have proved among the most impressive consequences of recent climate change in the UK. Flying insects, in particular, have paved the way and rendered obsolete a fair number of old distribution maps. Increasingly, via immigration from the near continent, they have also added to the UK species lists of regular visitors and established breeders. I find it amazing that the fragile, rice-paper wings of butterflies and moths regularly survive carriage on the wind for hundreds, even thousands, of kilometres. In extreme cases, insect mass migrations are awe-inspiring. Perhaps best known in the UK is the arrival every summer of Painted Ladies *Vanessa cardui*, sometimes in thousands or even millions, all originating from around the Mediterranean. The journey can be accomplished in a matter of days, at altitudes as high as a kilometre, when the wind is in the right direction. In the autumn they head back south again, but many undoubtedly perish on these perilous journeys. Rarer but even more outrageous trips are occasionally made by North American Monarch butterflies *Danaus plexippus* that somehow survive a one-way voyage across the Atlantic and turn up in the Isles of Scilly.

Painted Ladies are welcome summer invaders, travelling huge distances en masse to reach British shores. They exemplify the dispersal capabilities of seemingly fragile flying insects.

Unlike the situation with vascular plants, where about half our species are not native to the UK and many have been moved around the country by humans, most of our invertebrates are native and few are moved deliberately from one place to another by people. Nevertheless, some newcomers did not arrive of their own accord, and when attributing changes to climate effects we must keep a wary eye out for them. There are some well-documented examples. The New Zealand Flatworm *Arthurdendyus triangulatus* certainly didn't travel halfway round the world under its own steam, and could not in any case cross a sea unaided. First seen in Northern Ireland in the 1960s, it continues to spread but has not devastated earthworm populations as badly as initially feared. It probably arrived in topsoil with imported plants. More problematic have been Zebra Mussels *Dreissena polymorpha*, originally native to Russia but now rampant in waterways across Europe and North America. They probably came attached to ships and have been in the UK for some 200 years. Chinese Mitten Crabs *Eriocheir sinensis* hail, as their name suggests, from east Asia and have taken particularly well to the Thames estuary, where on one occasion, according to the Conservationjobs website, they were watched 'leaving the river and moving towards Greenwich High Street'. But perhaps the worst example of an invasive alien invertebrate damaging British wildlife is the North American Signal Crayfish *Pacifastacus leniusculus*, released from farms into British rivers in 1976 and still spreading like wildfire, killing off our native White-clawed Crayfish *Austropotamobius pallipes* as it goes. The crabs and the crayfish are highly palatable, and one trapper collected 100 tonnes of *Pacifastacus* in a single year. Unfortunately, this huge quantity signifies the impossibility of controlling the invasion rather than suggesting a practical solution for stopping it.

These introductions are readily attributable to human activity, but the status of some species is less clear-cut. Swallowtail butterflies are indisputably native to parts of the Norfolk Broads, but only in that very limited region of Britain. In 2014 a dozen or so adults appeared in Sussex and bred successfully, at least producing caterpillars, though there is no evidence yet that a new population has established. Was this heralding a new invasion from the continent? Hopefully yes, but this spectacular insect is bred in captivity by fanciers, and escapes from such a colony cannot be ruled out. There are similar concerns about at least one other impressive butterfly, the Large Tortoiseshell *Nymphalis polychloros*, once widespread in the UK but now rarely seen and, when

Mole Crickets are an enigmatic species that were once common in southern England but for which new records are always suspected as accidental introductions.

it is, also likely to be an escapee from a breeding facility. But perhaps the most frustrating insect that falls into this category of uncertainty is the Mole Cricket *Gryllotalpa gryllotalpa*. A noisy inhabitant of damp meadows, it was once widespread in southern Britain but is now essentially absent, unless recent records from the New Forest turn out to be a viable population. In Gilbert White's day Mole Crickets were not uncommon around Selborne, and 'As they often infest gardens by the side of canals, they are unwelcome guests to the gardener ... and occasion great damage among the plants and roots, by destroying whole beds of cabbages, young legumes and flowers.' Every year there seems to be a new record of a Mole Cricket somewhere in the UK, but invariably these have been traced back to soil accompanying plant imports. Gardeners to blame again, albeit inadvertently, confounding the issue as to whether our native crickets still persist somewhere. It would be truly wonderful if they did, but for now we have to wait for more information from the New Forest. Elsewhere such hopes have long since faded.

In looking at changes in distributions of invertebrates and attributing them to climate change we must therefore be cautious about possible human interference, but in general we are on safer ground than with plants. The signals to look for are broadly the same, though: movement of northern or, more rarely, southern range borders, because some invertebrates are adapted to cool environments, and infilling of pre-existing ranges. As with plants, widespread habitat deterioration over recent decades means that range expansion or increases in abundance will rarely be due to habitat improvements.

Lepidoptera: butterflies and moths

Phenology

As documented by Burton & Sparks (2002), it was evident by the turn of the millennium that butterflies were responding similarly to plants in the sense that flowers in woods and fields, and fluttering wings on early spring days, were all appearing earlier than in the recent past. The insect activity changes were sometimes dramatic, very much along the lines of those seen with spring flowers.

All 31 butterfly species with records from southern England were flying earlier in the 1990s than in the 1970s, but the degree of change depended on life-cycle. Excluding Red Admirals, which only started overwintering in England recently, differences in first sightings of species that hibernate as adults were by far the largest, averaging a remarkable two months earlier than in earlier times. Peacocks were typically seen towards the end of March in the 1970s, but in the 1990s it was by mid-January. Brimstones and Small Tortoiseshells, among others, were now worth looking out for at the start of February or even in January, compared with sometime in March a couple of decades earlier. Next in line were species overwintering as pupae, such as Orange Tips, Large Whites, Small Whites and Holly Blues.

Red Admirals mostly arrive as summer migrants, but increasing numbers are overwintering successfully in the UK.

From mid-April first appearances in the 1970s, this group had moved on in the 1990s by an average of one month to around mid-March. Finally, butterflies overwintering as eggs or larvae showed the least, but nevertheless significant, trends to earliness, averaging changes of two to three weeks. More than half of the 31 recorded species were in this group, including the Common Blue *Polyommatus icarus*, Gatekeeper *Pyronia tithorus*, Grayling *Hipparchia semele* and Ringlet *Aphantopus hyperantus*. These are not early spring fliers, and all of them have to complete pupation or, in some cases, larval development before appearing on the wing. Nevertheless, from typical 1970s first sightings in mid-May through to mid-July, by the 1990s many were seen by early May or even late April.

Increasing spring warmth is the obvious, intuitive explanation for luring butterflies out of hibernation earlier than they appeared in past decades. Just as was true with spring flowers, correlation with climatic factors, especially temperature, is the only realistic way of testing this belief. For 27 of the 31 species investigated, such correlations were readily demonstrable. Mean February–April temperatures accounted for emergence dates of species including Holly Blue and Dingy Skipper *Erynnis tages*; in both cases emergence was about ten days earlier for each degree Celsius of warming in that early spring period. However, there was more to it than that. Analysing the vast data set now available from the UKBMS, from 1,622 localities between 1975 and 2012, Roy *et al.* (2015) found that there was greater consistency in butterfly flight times across the UK than would be predicted if all populations were responding to climate in exactly the same way. Overall, adult emergence dates became earlier in warmer years by an average of 6.4 days per degree Celsius, but mean differences among populations averaged only 4.3 days per degree Celsius. Emergence dates were therefore more synchronised over their geographic range than would be expected taking account of the differences in temperature regimes across the country. This was best explained by local adaptation to climate, genetic effects upon which plastic responses to climatic variation between years are superimposed. Plastic responses are those that can be made by individuals within their lifetimes; callusing of skin on human hands due to regular manual work is a classic example.

The consequences of this realisation are disconcerting, because it means that insects might be limited by local genetic constraints in their ability to respond to future climate warming. As mentioned earlier, distinguishing the influences of nature and nurture is an increasingly common theme in phenological research. Sometimes, as with the butterflies, their contributions can be disentangled by statistical analyses of large data sets obtained in the field, but otherwise laboratory studies, where environments can be carefully controlled, are needed to distinguish population genetic from individual plasticity responses.

Sometimes scientific analysis of phenology seems impersonal, looking at huge amounts of information gathered by a multitude of contributors. As an antidote to statistical exactitude it is nice to take a closer look at local observations, such as those of local enthusiast Harry Clarke and his colleagues in Surrey. National trends are reflected in the work of these naturalists, which includes observations on species not often registered elsewhere. It is especially fascinating to learn that the Silver-studded Blue *Plebejus argus*, a late flier, is one of the strongest responders to climate change in Surrey. *Plebejus* larvae are associated with ants; in this situation, phenological advances must somehow have involved a synchronous change for both species. Ants harbour the larvae of blue butterflies in their nests, and the Large Blue *Phengaris arion* became temporarily extinct in the UK when access to its obligate host ant was lost due to changes in habitat management. Apparently Silver-studded Blues have appeared between two and three weeks earlier in the 2000s compared with what was usual in the 1990s. I associate this pretty little insect with the Wealden heaths, and especially Woolmer Forest, where in a good year swarms of them are a truly inspiring sight on a warm summer's day.

One of Rothamsted's flying insect samplers, part of a network that among other things has revealed ongoing declines of many moths in England.

The 70 or so butterfly species native to the British Isles are vastly outnumbered by our 2,500 moths, some two-thirds of which are 'micro-moths' with wingspans of less than 2cm. This great diversity, the difficulties of identification, and of course the nocturnal habits of most species mean that studying them is something of a specialist subject. Few of us non-specialists are familiar with more than a handful of species, such as the hawk and tiger moths. Pleasure at finding a couple of hundred moths in my overnight light trap is rapidly tempered by the realisation that it will take many hours to figure out what they all are. Fortunately, there are increasing numbers of moth enthusiasts far more skilful than me.

Quite a few Victorian naturalists included moths as well as butterflies in their interests, and their work has been continued to the present day. Especially valuable is the network of more than 400 moth traps across the UK, operated as part of the Rothamsted Insect Survey (RIS) since the 1960s. Butterfly Conservation set up a National Moth Recording Scheme in 2008, now with contributors in the thousands, and older data sets have been retrieved for comparison. So, although perhaps not an ideal group for a 'citizen science' approach, there is plenty of information about moths out there.

As with so many butterflies, our nocturnal lepidopterans have often shown trends towards earlier flight times. In Scotland, Sparks *et al.* (2006) found that, between 1973 and 2003, 17 of 21 moth species were taken in traps between one (first catch) and three days (mean catch day of the year) earlier per decade. Once again there were correlations with climatic factors, mostly with temperature but sometimes also with rainfall in preceding months. Unlike butterflies, where most of the big changes were seen in early-spring emergents, the majority of moths fly in the summer months, and this complicates identification of crucial climatic variables. Among the species showing the strongest trends to earliness were the Common Quaker *Orthosia cerasi*, Common Marbled Carpet *Dysstroma truncata* and Hebrew Character *Orthosia gothica*. It has been pretty much the same story in Ireland, where O'Neill *et al.* (2012) discovered that 44 out of 59 moth species were flying progressively earlier between 1974 and 2009. Most of these species were also on the wing for longer as the years went by, and, as with the butterflies, it was the first fliers that showed the largest trends towards earliness. And, yet again, there were correlations with temperature increases, especially in May and June. Interestingly, as crops up in virtually all groups when phenology is examined, there are mavericks trending in the opposite direction to the majority. In Scotland, the Garden Carpet *Xanthorhoe fluctuata* was emerging about 20 days later, on average, in 2003 compared with 1973. This is a species of midsummer flight and, like all such anomalies, surely warrants special attention. Exceptions to a rule can often provide unparalleled insights into what the rule really is.

As well as early appearances, another potential consequence of climate change that is beginning to emerge is variation in the number of generations per year of butterflies and moths (known as voltinism) with latitude. This is something that is often captured in the species accounts in field guides, and has been studied elsewhere in Europe by Altermatt (2010).

Distribution and abundance

Most British butterflies, more than 70 per cent of our species, are in long-term decline and have been in this sad state for decades. That is the unequivocal conclusion from the UK Butterfly Monitoring Scheme. Indeed, butterflies are faring among the worst of all animal groups and make a mockery of government promises to halt the UK's trend of biodiversity decline, first by 2010 and then, equally unrealistically, by 2020. Only landscape-scale changes in farming practices could achieve that goal, and although the piecemeal application of Countryside Stewardship schemes has been beneficial, it is unlikely by itself to be anything like enough.

Despite the gloom, a few butterflies have bucked the downward trend. In some cases declines have been arrested or even reversed by focused habitat management. Of these, the successful reintroduction of the Large Blue, following its extinction in 1979, is a striking example. This is one of several butterflies associated with grazed chalk grassland, especially south-facing slopes, and other habitat specialists in this group have benefited from a combination of good habitat management and, probably, a warming environment. The gorgeous Adonis Blue *Polyommatus bellargus* has increased in numbers and widened its distribution, mainly by range infilling, in southern and south-east England in recent years. Similarly, the Silver-spotted Skipper *Hesperia comma* has regained lost ground in the same region but is nowhere near to restoring its historical distribution as far north as Lincolnshire. Still, the omens for at least some species look good, and they are even better for a few generalists that require only widespread larval food plants such as Stinging Nettles *Urtica dioica*, woodland or meadow grasses, Holly or Ivy. In most of the UK these plants are ready and waiting for temperature-limited butterflies to arrive, and that is exactly what has been happening. The Speckled Wood, until recently limited to latitudes below Cumbria and Yorkshire, has advanced elsewhere in north-east England and invaded Scotland. The Ringlet, Marbled White *Melanargia galathea*, Gatekeeper and Comma have also been marching steadily northwards, while the Holly Blue and Small Skipper *Thymelicus sylvestris* recently turned up in Scotland.

The history of the Comma in the UK is particularly interesting. In early Victorian times this butterfly was found as far north as Scotland, but during the 19th century it declined enormously, to the point where Commas survived mainly in the Welsh borders. Since 1960 the

Comma butterflies are widening their distribution by mechanisms that include broadening the feeding niche of their larvae.

species has regained much lost ground and is now widespread again, also recolonising the Isle of Man, but if this increase is due in any way to climate change the mechanism must be more complex than just benefiting from extra warmth. Before its decline, hops were the main food plant of Comma caterpillars, and the butterfly's rapid disappearance from much of England was attributed to a concurrent reduction in hop growing. However, the Comma's resurgence has nothing to do with hops but has been coincident with a change of preferred food plant. Nettles are now used, and of course are common everywhere. This may have involved a genetic change permitting the switch, complicating any simplistic interpretation of range expansion due to climate. However, some other butterflies have also benefited from a widening of the larval diet, which may in some way be due to rising temperatures. The Brown Argus *Aricia agestis* has improved its lot in England by feeding more on various types of geraniums, among which are common garden plants in a habitat relatively undamaged by agrochemicals. Its recent northward expansion, averaging some 3km per year, is among the fastest of any British butterfly species.

Scotland is witnessing significant changes in its butterfly populations, many of which can be linked to climate because the effects of agricultural intensification are less overwhelming there than they are in much of England. Apart from the appearance of new species, as mentioned above, others such as the Peacock and

Orange Tip have enjoyed substantial population increases. But this country is also home to a few northern butterflies at their southern range limits in the UK, adapted to cool climes and potentially vulnerable to a warming effect. One such, the Northern Brown Argus *Aricia artaxerxes*, has certainly declined – though it is not clear whether this is related to climate change – while two other highland butterflies, the Scotch Argus *Erebia aethiops* and Large Heath *Coenonympha tullia*, seem to be doing OK in Scotland, at least for now. It is notable, though, that the Large Heath has virtually disappeared from most of its old haunts south of the border, and local extinctions of Northern Brown Argus and Scotch Argus have been most frequent in southern locations (Gillingham *et al.* 2015).

Continent-scale studies of climate effects on butterflies in North America and Europe have shown that whereas many species have extended northwards around their range edges in British Columbia and in Scandinavia, there have also been contractions northwards at southern limits in California and Spain. Lag periods occur between changes in the climate envelopes and subsequent changes in distribution, but these tend to be longer at the southerly, retreating border compared with at the northerly, expanding one. In the UK, just as with arctic-alpine plants, there is not much scope for cold-adapted butterflies to head further north. There is certainly no room for complacency about the future prospects of our cold-adapted butterflies.

While no totally new species of butterflies have arrived to stay in the UK in recent decades, there have been changes in the incidence

The Northern Brown Argus has declined across its range in northern Britain, and climate change may be a contributory cause, though habitat loss has also been substantial.

of some existing natives that may foretell the future. We have seen the arrival of Swallowtails and their successful reproduction on England's south coast, and the increasing overwinter survival and occasional reproduction of Red Admirals and Clouded Yellows *Colias croceus* may preface a change from regular visitors to resident breeders in the not-too-distant future.

The question arises as to how exactly a warming climate affects butterflies. The evidence suggests that buoyant, thriving populations are a prerequisite for range expansion, and small or declining ones are unlikely to spread however much the weather changes in their favour. This is in line with a general expectation of population ecology, that when reproductive success is high there will be pressure for some offspring to move beyond their birth area and thus reduce competition from siblings. Inevitably, though, there will be a lag between an area becoming climatically suitable and a successful colonisation event. It takes time, and maybe an outstandingly successful reproduction year, to find that tantalising new habitat patch. The existence of such a lag was demonstrated convincingly by the deliberate movement of Marbled Whites and Small Skippers to two sites in northern England tens of kilometres beyond their existing range boundaries, but within a climate envelope estimated now to be suitable for them. Both species survived, bred and subsequently thrived. This experiment is one of many contributions to our understanding of how climate change is affecting British wildlife by Chris Thomas and his colleagues at the University of York, together with Stephen Willis and Brian Huntley at Durham University. The approach demonstrates how scientific studies can move conservation along.

Another recent climate beneficiary, the Speckled Wood, has provided further insights into how warming works its magic. As its name suggests, this species has previously been thought of as a woodland butterfly, but over the past 40 years Speckled Woods have become less restricted to this habitat in places where temperature and summer rainfall have increased the most. As demonstrated by Pateman *et al.* (2016), using a combination of field observation and both field and laboratory experiments, open habitats of the type now occupied previously supported low larval growth rates. Compared with woodland, they were cooler in winter and more prone to drying out, and thus host plant deterioration, in summer. Presumably this combination was lethal to the insects. What climate change has done is to broaden the niche of the Speckled Wood and thus pave the way for its recent range extension. Given the changes in food-plant selection mentioned above for the Comma and the Brown Argus, this may be a common mechanism at work under current climate change.

And what of the UK's moths? The main story here, as with butterflies, is one of catastrophic decline in southern and central England. Further north, and in Scotland, moths are faring much better, and once again the blame for decreases in the English lowlands can justifiably be directed at intensive farming with its extravagant use of fertilisers and pesticides. The RIS has detected an overall decline in moth abundance of about 40 per cent in southern England, and of more like two-thirds in the larger species, during the last 40 years. No fewer than 65 species have gone extinct during the 20th and 21st centuries, and several more are on the brink. These changes have been all too obvious to those of us old enough to have been driving cars since the 1960s. In those days a night-time trip in summer would invariably end up with a windscreen sticky with dead moths, and the task of clearing up what windscreen washers manifestly failed to do properly. How we long for those days now, with the 'moth snowstorm' just a distant memory.

Superimposed on habitat effects, just as with butterflies, is the influence of climate change. Again there have been winners and losers. Overall, between 1970 and 2010, 260 moth species declined while 160 increased. Once again there were cases of range expansion by warmth-lovers, and contractions by cold-selected species, as enumerated by Fox *et al*. (2014). The striking Jersey Tiger *Eupalagia quadripunctaria*, until recently confined to the far south-west of England, has spread extensively eastwards and northwards. It is now regularly met with in my Somerset home village, flying by day and a real attention seeker. By contrast, the Garden Tiger *Arctia caja*, a common insect almost everywhere in my youth, has declined in England by over 90 per cent. This seems to be a climate effect, as it remains abundant in Scotland and in Somerset can still be found high up on the Mendips, but not often lower down. Up country there is a greater

BELOW: The Garden Tiger, a fast-declining moth that seems to be suffering from warmer temperatures.

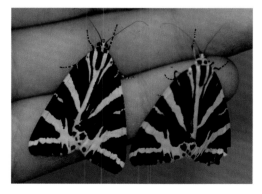

BOTTOM: The Jersey Tiger, a moth that is extending its range in southern England in response to climate change.

preponderance of success stories, though not always with clear-cut causes. Moth expert John Knowler has kept details of moth species that have moved north into central Scotland since 2000:

*They are, in order of appearance, the Red-necked Footman (*Atolmis rubricollis*), Slender Brindle (*Apamea scolopacina*), the attractive micro* Ypsolopha sequella, *Alder Moth (*Acronicta alni*), Copper Underwing (*Amphipyra pyramidea*), Oak-tree Pug (*Eupithecia dodoneata*), Pale Pinion (*Lithophane hepatica*), Buff Footman (*Eilema depressa*), Beautiful Snout (*Hypena crassalis*), Southern Wainscot (*Mythimna straminea*), White-pinion Spotted (*Lomographa bimaculata*) and perhaps Lesser Treble-bar (*Aplocera plagiata*) although this species may have been previously overlooked.*

But he also warns that:

With all of these species moving north, it is tempting to see their range expansion as a response to climate warming, but the factors influencing whether a given species can expand its range must be more complex. The movement of the footman species must be influenced by cleaner air that supports the growth of the lichen that the larvae feed on, but why has Buff Footman made it to central Scotland when the closely related species, Dingy and Scarce Footman, both occurred further north than Buff Footman when mapped in 1978 but they have not spread to central Scotland? The larval food plant of Slender Brindle is described as the woodland grasses Wood Melick, Wood Meadow Grass and False Brome. These grasses are rare or absent in the acid woodlands of western Scotland and yet I have caught up to 13 Slender Brindle in a single trap. Clearly, its move north must have been supported by the adoption of a more catholic range of foodplants.

Yet again, we have the possibility of widening niche breadths under the influence of climate warming but also complications of interpretation where other, often unknown, factors are probably in play.

Despite the gloom and recent extinctions, there are some recent moth newcomers in England, most of which probably crossed the English Channel without help from humans. At least 27 new species have become established and many more recorded as occasional visitors. Climate warming has almost certainly underpinned these invasions. One such recent arrival, the Horse-chestnut Leaf Miner *Cameraria ohridella*, was first seen in England in 2002 after a rapid cross-continent expansion from its Balkan heartland, followed by a similarly quick spread across most of England. This moth, which blights but does not kill Horse-chestnut trees, is perhaps one of the less welcome new colonists to reach our shores. Most, if not all of the other new arrivals are harmless by comparison as far as we know, and can therefore be welcomed.

Newer perspectives

Until recently, there has been a widespread belief that climate warming has on the whole been beneficial, at least to most lowland butterflies and moths. However, recent analyses of trends in multiple species have shown that the benefits are far from ubiquitous, and it is not only butterflies at their southern range limits that are being adversely affected. Declines in about a third of lowland butterfly species and in half of the moths investigated have probably been exacerbated by climate change over the past 40 years, a startling and very disturbing result that surprised the optimists among us, basking smugly in the good news about range expansions. Among the previously unsuspected victims are the Wall butterfly *Lassiommata megera* and moths such as the Dark-barred Twin-spot Carpet *Xanthorhoe ferrugata*, Mouse Moth *Amphipyra tragopoginis* and September Thorn *Ennomos erosaria*. It seems that new and unfavourable combinations of temperature and rainfall at critical times of year have contributed to decreases in these species, a pattern of cause and effect not obvious to a casual observer and only unravelled by clever statistical analyses. So even the good news has not turned out to be as universal as was first thought.

Odonata: dragonflies and damselflies

Phenology

Second only to butterflies in their conspicuous daytime activities, here we have another group of insects that are demonstrably responding to climate change. The UK has just 40 or so species, but this low diversity is more than made up for by the charisma of our delicate, fairy-winged damsels and macho, attack-helicopter hawker dragons. None of them overwinter as adults, but always as eggs or larvae in ponds, ditches, rivers or streams. Sometime in spring or summer the aquatic larvae crawl out of the water, usually up emergent vegetation, and metamorphose into adults that subsequently patrol their birthplaces and, sometimes, far beyond. This engaging group of insects is gaining ever more popularity among naturalists, and the British Dragonfly Society runs a national survey and monitoring scheme.

Hassall *et al.* (2007) revealed that between 1960 and 2004, dragonflies and damselflies generally appeared on the wing seven days earlier, on average, at the end of this recording period compared with at the start of it. Overall length of the flight periods did not change, however – which is not easy to explain, especially as autumn cooling has become later and the temperature regimes over which flight is possible are certainly longer than they used to be. Adult damsels and dragons are killed by frost if they live long enough to encounter one, and none make it through the winter. Perhaps flight period is dictated by individual survivorship and the chance of being predated, which probably hasn't changed, rather than adequate warmth. Species with egg diapause, in which eggs laid the previous year sit out the winter and hatch in spring, were less responsive to the ongoing warming than those with larvae active in the previous summer and which overwinter and metamorphose the following year. Dragonflies demonstrate that aquatic habitats have been warming up sufficiently, mostly in winter and early spring, to impact on phenology in a similar manner to what has happened with terrestrial species. We might expect temperature fluctuations in water to be dampened relative to air, but evidently that effect has not been strong enough to suppress a phenological response. Spring fliers changed their phenology more markedly than summer ones, probably because among the latter group egg diapause was more common. The strongest responders overall were the Emperor Dragonfly, the Keeled Skimmer *Orthetrum coerulescens* and the Large Red Damselfly *Pyrrhosoma nymphula*, all spring species with overwintering larvae.

Distribution and abundance

Anyone who has been 'buzzed' by a Southern Hawker *Aeshna cyanea* will be well aware of the impressive flying power of the larger dragonflies. Even the fluttery flight of damselflies regularly takes them far from the ponds of their birth, and perhaps it is no surprise that this group of insects has made the most dramatic response to climate change in terms of distances covered to colonise new sites. Some of the changes in dragonfly distributions over recent decades have been quite astonishing, as Brooks *et al.* (2007) have shown. Steve Brooks reflects on a rapidly changing scene:

> *When I first became interested in dragonflies in the 1970s we still had incomplete knowledge about their distribution in Britain. However, any changes in the fauna appeared to be driven by human-induced land-use changes, resulting in declines and even extinction of species. Now, thanks to the efforts of thousands of dragonfly recorders, we know that over the last 20 years at least 70 per cent of our dragonfly species have been shifting their ranges northwards and westwards in response to climatic warming and improvements in the quality of our rivers. Where I grew up in the West Midlands it is now possible to see many more species than in the recent past due to the northern movement of resident species and the increasing number and abundance of migrant species. We are now even seeing species beginning to colonise Britain that have never been recorded here before, including one species that became extinct in 1953 and has returned to the east Thames marshes.*

A Southern Hawker dragonfly freshly metamorphosed from its larval skin. Long larval development is one constraint on the dispersal rates of damselflies and dragonflies.

Records collected by, among others, members of the British Dragonfly Society reveal that between the periods 1960–1970 and post-1985, all but three of the 37 resident species of dragonflies and damselflies extended their northern range edge northwards by an average of 74km, much further than the most mobile butterflies. Big winners invading northern England recently include Migrant Hawkers *Aeshna mixta*, Southern Hawkers, Emperors (see map overleaf), Black-tailed Skimmers *Orthetrum cancellatum* and Ruddy Darters *Sympetrum sanguineum*. All are habitat generalists, common in southern

The Ruddy Darter has raced northwards as far as Scotland as a result of climate change.

An Emperor Dragonfly, one of several species that is flying earlier in summer and which has extended its northern range boundary in the UK.

and central England and found in a wide range of water bodies. These and other expansions have resulted in novel appearances of Migrant Hawkers and Emperors in Scotland and Ireland, and of these species together with Brown Hawkers *Aeshna grandis*, Broad-bodied Chasers *Libellula depressa*, Black-tailed Skimmers, Ruddy Darters and Banded Demoiselles *Calopteryx splendens* in Scotland. The Hairy Dragonfly *Brachytron pratense* has become much commoner in northern England. Experienced naturalist Brian Banks recalls:

> *A sign of the times. When I started working at Sussex the Hairy Dragonfly was then a coastal grazing marsh scarce species, stretched along the south coast, and just getting around the coast of East Anglia. My sister described one in Durham the other day, and on checking the NBN atlas they are indeed found all the way up to the north-east and west Scotland and are in just about every 10km square down here. I only saw one dragonfly ever when I lived up there. Global warming must be taking off if those beasts can survive up there.*

OPPOSITE PAGE:
Changes in the distribution of the Emperor Dragonfly. Adapted from Parr (2010).

Unlike the habitats of most terrestrial insects, some of the freshwater habitats used for breeding by dragonflies have improved in quality since the 1970s. Victorian abominations, the industrial graveyards of the UK's vibrant rivers, are increasingly being consigned to history. Does this good news compromise interpretation of dragonfly range

Invertebrate tales

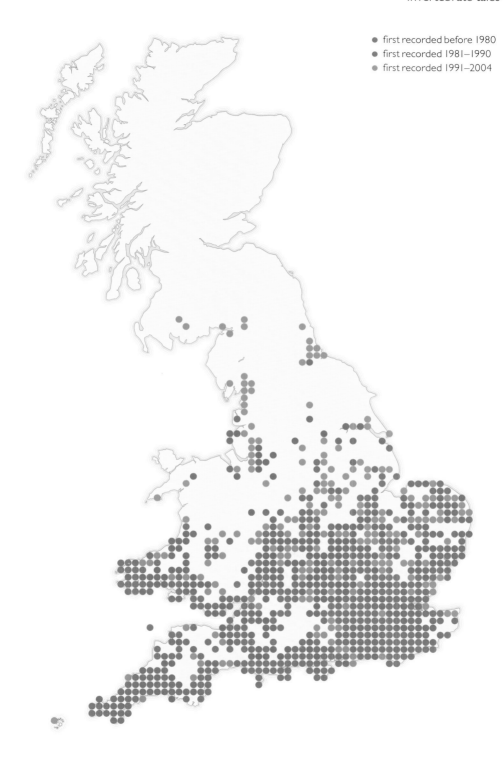

- first recorded before 1980
- first recorded 1981–1990
- first recorded 1991–2004

The Small Red-eyed Damselfly is one of the UK's new colonists, now well established and spreading ever more widely in England.

expansions? Probably not in most cases, because the main beneficiaries of better water quality have been streams and rivers, rather than the ponds and ditches that are home to the majority of British dragons and damsels. These include species with the most dramatic range expansions. Ponds have not, unfortunately, improved their lot in parallel with rivers. For over a century, ponds declined in both number and quality, although conservationists including the Freshwater Habitats Trust have had some recent success in moderating these trends. However, reduced pollution may have helped the revival of some riverine species such as Banded Demoiselles and White-legged Damselflies *Platycnemis pennipes*, which have experienced substantial range infilling as well as northward range extensions.

Another pleasing development has been the increased frequency of visits by European dragonflies and damselflies not previously resident in the UK, in some cases followed by breeding successes. Foremost among the newcomers is the Small Red-eyed Damselfly *Erythromma viridulum*. First seen in Essex in 1999, it has spread widely since then in east and south-east England and is surely here to stay. Another even more recent invader is the Willow Emerald Damselfly *Chalcolestes viridis*, now well established in parts of southern England. Other species not yet securely entrenched as new residents, but ever more frequent visitors from the continent, are the Southern Emerald Damselfly *Lestes barbarus*, Lesser Emperor *Anax parthenope*, Scarlet

The White-faced Darter is under threat from climate change, with a southern range edge recently moving northwards.

Darter *Crocothemis erythraea* and Banded Darter *Sympetrum pedomontanum*. These, too, make increasingly successful breeding attempts – as has the Red-veined Darter *Sympetrum fonscolombii*, an occasional visitor for more than a century past.

It hasn't been entirely a story of riotous expansion. Four British species are essentially northern, and all have retreated further north as the climate has warmed up. Azure Hawkers *Aeshna caerulea*, Northern Emeralds *Somatochlora arctica*, Northern Damselflies *Coenagrion hastulatum* and White-faced Darters *Leucorrhinia dubia* have, however, found space for northern range extensions within Scotland, and this is probably happening, so far, without any overall range contraction. Of these species, the White-faced Darter is particularly interesting, because until the 1990s it maintained outposts as far south as Surrey. At that time it was one of a famous contingent of Odonata at Thursley Common, one of the most diverse sites in the UK for this group of insects. Towards the end of their time at Thursley, White-faced Darters could be seen around one or two acid bog pools surrounded by boardwalks for easy viewing. Various explanations, including too much disturbance, were proposed for their demise, but in retrospect it looks like the place, an open heath, just became too warm for them. By the 2000s the most southerly locations for White-faced Darters were in the north Midlands, where conservation management has so far proved successful in maintaining viable populations.

All dragonflies and damselflies are sun worshippers, and higher temperatures constitute a good explanation for the recent increases in range, though we do not know whether particular, or perhaps all, life stages benefit from this change. With butterflies it seems that part of a climate response can involve niche broadening, but so far there is not much evidence of that among the Odonata. Perhaps we need to look harder at habitat selection, though, because the White-legged Damselfly now breeds in standing water rather than just in the slow-flowing rivers to which it was previously confined. This broader niche is similar to its habitat selection further south in Europe, so maybe other species are showing comparable changes in the UK.

There may be other factors at play too, because light-coloured butterflies and dragonflies are spreading north more than dark ones as the climate warms. Dark colour is an adaptation to cool climes, because its light-absorbing qualities allow animals to warm up quickly. Lighter-coloured creatures can only move north when there is less need for this benefit. But as with butterflies, it is important to be aware that despite climate-mediated range expansions about a third of British dragonflies and damselflies are in decline, primarily because the number and quality of countryside ponds continues to deteriorate in much of the country.

Hemiptera: bugs

Phenology

Astonishingly to my mind, a group of tiny bugs was among the first of the insect clan to reveal phenological change linked to climate. These are the aphids, the curse of farmers and gardeners everywhere and therefore of economic as well as ecological interest. Aphids proliferate enormously every summer, primarily by asexual reproduction, and can devastate crops and garden plants by sucking sap from them. Most individuals are sessile and stay on the host plant where they were born, but others, 'alates', grow wings and disperse to pastures new. Despite their bad press, I rather like aphids. Greenfly and blackfly constitute a wonderful, hyper-abundant food supply for freshly metamorphosed froglets and toadlets in captive vivaria. More importantly, they form a substantial part of the diet of House Martins *Delichon urbicum*, when the insects fly or are swept into the atmosphere by passing winds. House Martins have declined hugely in most of England, though interestingly not in Ireland or Scotland. Almost certainly the regional differences arise because the agrochemical industry has annihilated aphids, as well as many other invertebrates, in the intensively farmed parts of the UK. From an ecological perspective we need more, not fewer, of these important components of the food chain.

The Rothamsted Research Centre in Hertfordshire is the oldest agricultural research institution in the world and has, for over 50 years, organised surveys of a wide range of insects at sites across the UK on a regular basis (see also pp. 85–86). Aphids have long featured prominently in the RIS scheme. The insects are collected by suction traps suspended approximately 12m above ground, the optimal height for aphid flight. The main period for aphids historically started in April, but all 55 aphid species recorded by the RIS have started flying ever earlier in the year, and most are now also active for longer each summer than used to be the case, as described by Bell *et al.* (2015). Aphids require particular temperature thresholds to initiate flight, and weather has been a key factor in bringing flight times forward. Correlations of first flight times were strongly related to two factors: the strength of the North Atlantic Oscillation (NAO) in winter and accumulated days with temperatures exceeding 16°C later on.

The NAO is a complex phenomenon based on differences in sea-level air pressures between Portugal and Iceland, and its strength (that is, the size of this difference) varies between years. A strong NAO is essentially a surrogate measure of mildness and dampness in winter. Good news for aphids, therefore, is a warm wet winter followed by a mild spring. Overall, average first flight times of all aphid species have advanced by about 0.6 days per year since 1965, so by about a month in 50 years, but there has been considerable interspecific variation. The greenfly *Utamphorophora humboldti* holds the record, advancing by 2.7 days per annum, a truly enormous change also associated with a longer flight period and population increase. Mostly, though, the RIS results show little or no change in aphid population sizes over time (so perhaps House Martins need much more than an aphid food supply), implying that

Greenflies were among the first indicators of climate change, flying progressively earlier over recent years.

changes in phenology are not necessarily linked to changes in fitness, meaning characters affecting survival and reproduction. The RIS was established following the 1962 publication of Rachel Carson's famous *Silent Spring*, at a time when farmers were using pesticides prophylactically and on lavish scales. Whether the intention was to reduce this kind of mass poisoning is not clear, but if so, with new generations of pesticides such as neonicotinoids still wreaking havoc in our countryside, it manifestly hasn't worked out.

Distribution and abundance

Within the British bug fauna a familiar pattern is repeated: many species show northward range extensions, though for this group of insects there is a particular need for caution because recording effort has been patchier than for the more popular taxa discussed above. For some species the increases have been dramatic, and for large, conspicuous ones the changes are almost certainly real. The Green Shieldbug *Palomena prasina* was formerly associated mostly with southern, especially coastal, counties of England, but it is now commonplace as far north as the Midlands. The population in my Somerset garden and the surrounding fields must be huge, judging by the numbers visible with minimal search effort in late summer. Another recurring pattern is range extension away from the coastal habitats to which some bugs were previously restricted. Examples of this trend since 1960 include the rhopalid bug *Chorosoma schillingi* and the leafhopper *Athysanus argentarius*, but there are more recent expansions too, including the shieldbug *Odontoscelis lineola* and the squashbug *Arenocoris falleni*. Arguably only the most recent examples are readily explicable by climate change, a dichotomy that reinforces the need for caution when attributing the causes of range expansions for inconsistently recorded animals.

Any national picture of distribution changes is of course dependent on reports from local areas around the country and the enthusiasm of individual recorders. Local enthusiast Ray Barnett has noted, over 20 years, the appearances of insects around Bristol that were not previously known there. Newcomers included the Box Bug *Gonocerus acuteangulatus*, for a long time known only from Box Hill in Surrey where, as its name suggests, it fed exclusively on Box *Buxus sempervirens* trees. Now it has broadened its horizons to include Hawthorn, Blackthorn, Yew *Taxus baccata* and plum trees

The advance of Box Bugs across southern England, a species previously confined to Box Hill in Surrey, has been particularly dramatic.

(*Prunus* species) in its diet. The Plane Tree Bug *Arocatus longiceps* was first seen in 2007 and quickly became widespread, and the Cinnamon Bug *Corizus hyoscyami*, once confined to the south coast, also arrived in the Bristol area, maybe en route to Yorkshire where it is also now established. Keen observers make range changes come alive.

Aside from distributional changes of native species, there has been an impressive wave of invasions, with over 50 new species added to the British bug list since 1990. It is likely that most of these arrived by assisted passage, among plants and soil imported for gardening or forestry purposes. However, some almost certainly got here under their own steam, flying or being carried by the wind across the English Channel from their nearest locations in mainland Europe. These include freshwater species such as *Naucoris maculatus* and the water boatmen *Sigara iactans* and *S. longipalpis*, for which human participation in their journeys seems unlikely. According to Stewart & Kirby (2010), the bug fauna of southern England is probably now more diverse than at any time in recent history, or possibly ever, as a result of these arrivals.

Orthoptera: grasshoppers and crickets

Distribution and abundance

Stirring sounds of summer, the stridulations of grasshoppers and crickets enliven many a countryside walk – although nobody seems to have investigated possible changes in the phenology of these choruses. This group of insects does, however, include some of the big northward movers in response to climate change. Somehow they have managed this with rather limited powers of flight, although as with the aphids there is variation among individuals in this capacity. Brachypterous adults (those with vestigial wings) dominate most populations of Roesel's Bush-cricket *Metrioptera roeselii*, but macropterous individuals (those with longer, functional wings) also occur and increase in frequency during periods of population growth and range expansion. A similar distinction occurs with Long-winged Coneheads *Conocephalus discolor*, though in this case the difference is between long-winged and very long-winged forms. Swarming and solitary phases are commonplace in the Orthoptera, with frequencies that can change suddenly, and dramatically, in response to alterations in the insects' environment.

Locust swarms are the classic example of this astonishing plasticity, but in the UK its manifestation on a smaller scale is the key to range expansions of some of our native species in recent decades.

The Long-winged Conehead is another beneficiary of climate change, heading north at an astonishing rate.

Roesel's Bush-cricket was found only in the Thames and Solent estuaries in the mid-20th century, while Long-winged Coneheads were confined to the south coast of England before 1980. Both species have since spread extensively in southern and Midland regions of the country. They are not alone. The Short-winged Conehead *Conocephalus dorsalis* has reached the north-west coast of England and Scotland, while the Speckled Bush-cricket *Leptophyes punctatissima* and the Slender Groundhopper *Tetrix subulata* have also expanded their ranges northwards. The Woodland Grasshopper *Omocestus rufipes* and the Lesser Marsh Grasshopper *Chorthippus albomarginatus* have extended their distributions in England, while the Stripe-winged Grasshopper *Stenobothrus lineatus* has spread in East Anglia. A flightless newcomer that has been moving north in Europe and may have reached the UK as an accidental passenger with garden shrubs, the Southern Oak Bush-cricket *Meconema meridionale*, is now established at several locations in southern England and will probably continue to extend its range there.

Roesel's Bush-cricket is a climate change winner, also rapidly extending its range northwards.

In Essex, the population size of Roesel's Bush-cricket and numbers of the macropterous form capable of flight relate to summer warmth the previous year. Gardiner (2009) points out that this relationship constitutes circumstantial evidence that it is climate change that has promoted the spread of this insect. In addition, and like some butterflies, there is limited evidence of niche broadening in at least two species of orthopterans in the UK, possibly facilitated by a warming climate. The Grey Bush-cricket *Platycleis albopunctata* is an inhabitant of coastal dunes, cliffs and shingle banks in southern England and is now benefiting from ever-extending sea walls constructed to reduce flood risks. The (mostly) flightless Bog Bush-cricket *Metrioptera brachyptera*, a heathland bog species, may now be extending its niche beyond its primary habitat of heather and Purple Moor-grass *Molinia caerulea*. Yet again, though, the more important story is that many of the UK's crickets and grasshoppers are in long-term decline through habitat loss and degradation, and the benefits of climate change, for some species, are superimposed on this bleaker picture.

Hymenoptera: wasps, bees and ants

Phenology

Hymenopterans include the well-known social insects that exist as colonies with hundreds or thousands of individuals living in nests dominated by a single queen. However, for every social species there are many types of solitary bees and wasps in the British countryside. These include those that colonise 'bee hotels', an increasingly popular feature of wildlife-friendly gardens.

For the one species of hymenopteran on the *Nature's Calendar* website, the Common Wasp, there has been little evidence of phenological responses to climate change so far. However, observations in Poland, which included the German Wasp *Vespula germanica*, indicated no change in the timing of queen emergence from hibernation since the 1980s – though, interestingly, there was a change in the timing of worker appearance. This was happening earlier, and related positively to April temperatures and rainfall. German Wasps are abundant and widespread in much of England and need close scrutiny (look at their faces!) to distinguish them from Common Wasps. Whether this confuses recording is a moot point, but perhaps any phenological change, if it does occur, will affect both species similarly.

Another conspicuous group of flying insects is the bees, and although protected over winter by the actions of beekeepers, Honey Bees *Apis mellifera* may have suffered some effects from climate change. Disconnect between the timings of spring flowering and first worker flights can cause crises in food supply, and extreme weather events such as heavy rainfall can also be traumatic. However, unlike the situation with distributional change, there is little evidence of phenological responses in bees. Or is there? Bumblebees delight us as they buzz around our garden flowers in spring and summer but have become among the most endangered of our insect fauna, mostly due to the shocking losses of wildflower meadows that have proceeded almost unchecked since the Second World War. Gardens are now major sources of pollen and nectar for those species able to use this habitat. I now see worker and occasionally queen Garden Bumblebees *Bombus hortorum* throughout the winter months, grazing on winter-flowering Honeysuckle *Lonicera perclymenum*. The Bumblebee Conservation Trust reports that fully active winter colonies of this bee are now regularly recorded in the milder parts

German Wasps have a triangle of black spots on their face (top), whereas Common Wasps have an anchor-shaped mark (bottom).

of the UK. So at least some bees seem to be changing their lifestyle in a profound way, discarding hibernation altogether. It would be intriguing to learn more about how this happens. Are summer nests still dying off and being replaced by new ones, the previous norm, or are they just carrying on?

Distribution and abundance

Some social wasps have almost certainly benefited from climate warming in the UK. In my youth, the New Forest was allegedly the only place where European Hornets *Vespa crabro* could be found in the country, although that cannot have been the whole of the story. Gilbert White noted them more than once in his Selborne garden, where 'though few in number they make havoc among the nectarines'. In recent decades they have spread over much of south, central and eastern England. In Somerset they nest in local woods, and occasional workers visit my garden ponds to drink. These fine insects are not aggressive unless provoked, unlike the Asian Hornet *Vespa velutina* which recently arrived in southern England and is already the subject of eradication efforts. Beewolves *Philanthus triangulum*, a scourge of beekeepers due to their predatory habits, were formerly a rarity in the UK. Since 2000 these wasps have greatly expanded their range in England and flown as far north as Yorkshire. And there are other new colonists including

European Hornets were, until recently, a rare insect in southern England. They now occur widely across southern and eastern England and continue to extend their range.

the Saxon Wasp *Dolichovespula saxonica*, a recent arrival from Europe that is now widespread in southern and central England. We also have the large, occasionally aggressive Median Wasp *Dolichovespula media*, first seen in Sussex in 1980, which has spread rapidly across much of Britain as far north as Cumbria and Durham. Climate warming has, among its claims to fame, undoubtedly increased our opportunities for meeting new varieties of stinging insects.

Among the bumblebees there have been many catastrophic declines and even extinctions since the mid-20th century. Can some, at least, benefit from climate warming? Sadly, the consensus seems to be that bumblebees are likely to suffer extra pressure rather than relief as a result of climate change. Across Europe and North America, Kerr *et al.* (2015) detected a general trend for southern range boundaries of bumblebees to move north as the climate becomes too hot, but not for compensatory northward shifts at the northern range edges. Nobody knows why the bees have so far failed to extend their ranges northwards, but this lack of compensation means that ranges are being progressively squeezed, and contracting overall. It may be that bumblebees are disinclined to cross large tracts of unsuitable habitat to find meadows new, at least not within the timescale of rapid climate warming. The bees are not completely incapable of responding to a warming world, however. In some places they are moving to higher and therefore cooler altitudes, which offer at least a temporary respite.

The Bilberry Bumblebee is a cool-weather specialist with a southern range border contracting northwards.

Within the UK there is already at least one example of range contraction probably attributable to climate change. The Bilberry Bumblebee *Bombus monticola*, an upland species primarily of Scotland and northern England, seems to be retreating along its southern range margin. We do, on the other hand, have a new and very successful invader in the Tree Bumblebee *Bombus hypnorum*. First seen in Hampshire in 2001, this bee spread rapidly and reached Scotland within a decade. Although the initial colonisation might have followed climate warming and range extension in northern Europe, its subsequent expansion in the UK may reflect mainly the extent of suitable habitat available within its climate envelope.

Just occasionally, habitat change can assist rather than worsen the plight of bumblebees, and the extension of sea walls around much of the UK's coastline is acting as a corridor to reduce population isolation and allow extended movements of at least a few bee species. Sadly there are also downsides to sea walls, which create a barrier separating freshwater and saltmarshes much more sharply than before their creation (see Chapter 7). One such construct in Lancashire was, because of this effect, responsible for the extinction of that county's only Natterjack Toad population. It seems that no news is always good news.

Bumblebees are only a minor contributor with respect to total bee diversity. There are also more than 200 species of solitary bees in the UK, some of which are probably responding to climate change. The Dark Blood Bee *Sphecodes niger* is historically a southern English species that has extended its range substantially since 2000, as has the Yellow-legged Mining Bee *Andrena flavipes*. A particularly dramatic new addition to our fauna is the Violet Carpenter Bee *Xylocopa violacea*, one of Europe's largest bees that has been moving north across the continent for many years and was first seen in Wales in 2006. Since then it has become established as a breeding species, and it reached Northampton by 2010. It is hard to believe that warmer weather has not been a key player in the successes of these bees, though definitive evidence about any connection is lacking. Many of the species concerned are important pollinators, but it remains unclear as to whether, or by how much, they are benefiting from climate change rather than other changes in their environment.

For ants there is little to say, at least little that is sensible. We have been warned by the press that giant flying ants are invading the UK as a warming climate creates perfect breeding conditions. There may be trouble ahead if the invasive garden ant *Lasius neglectus* manages to cross the English Channel. This species has tiptoed north across Europe from around the Black Sea, and creates supercolonies in interconnected nests with several queens. The main cause of complaint is that they help aphids to proliferate, releasing sticky secretions that make a mess on parked cars. Sounds like a small price to pay if it indirectly helps House Martins.

Lasius neglectus is a widespread invasive pest in Europe, most likely originating from Asia Minor.

Other insects: flies and beetles

Distribution and abundance

It feels disrespectful, even perverse, to consider groups that together make up more than half of the UK's insect fauna in a single section. The fact is, though, that evidence of climate change effects on these animals is sparse. Recording schemes are continually improving, but much of the historical information for comparison with recent records is patchy and insufficient to infer trends over time. That's not to say there is no news at all. The impressive Hornet Mimic Hoverfly *Volucella zonaria* arrived in the UK from neighbouring Europe in the 1940s and until recently was only sighted near the south coast. Not so now: the fly has moved north fast since 2000 and reached Cheshire by 2008. Other hoverflies are also on the move. *Sphegina sibirica* moved west across Europe in recent times and since the 1990s has been regularly reported in the UK as far north as Scotland. Another species, *Epistrophe melanostoma*, also arrived in Britain recently and seems to be spreading. Fruit flies of the Tephritidae family (not the drosophilids familiar to lab geneticists) are also making news. The tiny *Tephritis cometa*, with its distinctively patterned wings, was known in the London area in the 1950s but has subsequently reached Scotland, and another equally pretty species, *Tephritis divisa*, reached southern England in the early 2000s and has spread extensively there since that time. It seems certain that other flies must also be doing better in warmer summers, but of course many will be suffering like most other insects from a disenfranchised countryside.

The Hornet Mimic Hoverfly has been in the UK since the 1940s, but only since the 1990s has it extended its range northwards.

On the other side of the equation, mayflies are experiencing adverse effects of climate change. Water temperatures in some British rivers have risen by up to 2°C or so over the past 20 years, with the result that the aquatic larvae of these flies now often mature after one year instead of two. Why does that matter? Partly because the emerging females are smaller and less fecund than they were in previous times when larval development routinely took two seasons, and partly because a two-year larval cycle allows an overlap of generations such that a disastrous breeding year can be recovered in the following one. That cannot happen in a single-year cycle. Anglers are noticing the difference, tying smaller flies than previously to catch trout. Clouds of Green Drake Mayflies *Ephemera danica*, once an almost universal spectacle on clean rivers in spring, are ever smaller and less often witnessed despite a general improvement in water quality. This makes temperature increase a prime suspect, at least as a contributor to mayfly declines. On top of that, the UK's only Arctic relict species, the Upland Summer Mayfly *Ameletus inopinatus*, is retreating to higher altitudes as its breeding streams warm up in northern England and Scotland.

Information about beetles is also sparse, considering the number of species living in the UK. The Harlequin Ladybird *Harmonia axyridis* and the Bryony Ladybird *Henosepilachna argus* are recent arrivals and both are widening their distributions, but only for the latter species is climate change a reasonable causative assumption. Harlequins are an Asian beetle that was first seen in the UK, in Essex, in 2004. Within a decade it had reached every corner of the country, earning the dubious accolade of the fastest-spreading invasive species yet seen. It seems

Bryony Ladybirds are a recent invader of the UK and are continuing a northward march that started in central Europe as the weather warmed.

The Screech Beetle is one of several water beetles extending their ranges northwards, in this case as far as Scotland.

unlikely, or at least unprovable, that climate change had anything to do with this success, which was probably down to an open, available niche already within an acceptable climate envelope. By contrast, the Bryony Ladybird was spreading northwards in Europe before its first sighting in the UK in 1997. It has since appeared at many locations, especially around London but as far north as the Midlands. Unlike the situation with Harlequins, the recent success of the Bryony is quite likely linked to the warming climate.

Water beetles are, perhaps surprisingly, a particularly mobile group of insects, especially those associated with still waters that can be prone to drying up. We have known for many decades that big beasts such as the Great Silver Water Beetle, and great diving beetles of the genus *Dytiscus*, are regular nocturnal fliers that sometimes alight on greenhouse roofs or appear in light traps set for moths. Within days of creating a new pond in my garden a few years ago, the water was colonised by dozens of the small diving beetle *Agabus bipustulatus* that must have flown at least a couple of kilometres from the nearest possible source. A recording scheme for water beetles started in 1979 and has provided indications of possible climate change effects, as documented by Aquatic Coleoptera Conservation Trust chairman Garth Foster in 2016. A few species seem to have arrived in south-east England from mainland Europe recently, including *Nebrioporus canaliculatus*, *Hydrovatus cuspidatus*, *Hygrotus nigrolineatus*, *Laccobius simulatrix* and *Limnebius crinifer*, of which only *Hydrovatus* and *Hygrotus* have subsequently extended their ranges northwards. Some longstanding natives have also headed up country. Since 2000, several have made their first appearances in Scotland, including the whirligig *Gyrinus urinator*, the Screech Beetle *Hygrobia hermanni*, *Ilybius chalconatus*, *Rhantus grapii*, *Dytiscus circumflexus*, *Enochrus melanocephalus* and *Cercyon sternalis*. While the majority of these clearly came from England, the distribution of *Rhantus grapii* suggests an invasion via Ireland and the Isle of Man. However, Foster emphasises that discovery of new immigrants in Scotland must be balanced by the admission that some supposedly relict species continue to be discovered, including *Hydroporus scalesianus* and *Ochthebius alpinus*, the latter not only recently in the field but also in museum material from 1953. Migration activity has also been detected within Scotland, with a few species newly reaching the Hebrides and/or extending up the east coast, in particular *Noterus clavicornis* and *Rhantus suturalis*.

I suspect that much more must be going on, especially among terrestrial beetles, that has as yet remained undetected. What is already clear is that not all beetles are benefiting from climate change, and a now familiar pattern is repeating itself with the Northern Dung Beetle *Agoliinus lapponum*. This insect of grazed uplands is widespread in Scotland and northern England. Comparing information collected in 2006–2008 with surveys in the 1950s, the lower elevation limit where this beetle still survives is more than 150m higher than it was in the mid-20th century. This change coincides with a mean annual temperature increase in its habitat zone of more than 1°C. *Agoliinus* is another species for which, thanks to a warming climate, the future does not look so good.

Other invertebrates

Distribution and abundance

Winged insects are well placed, on account of their mobility, to respond quickly to a changing environment, and it is no surprise that many of them are among the most impressive range expanders in recent times. However, they are not alone. Several spiders have increased their distributions coincident with climate change, and the question arises as to how some quite large range extensions have been accomplished by such seemingly sedentary animals. Geoff Oxford of the British Arachnological Society comments that 'there certainly are some spectacular northwards shifts in range for some spiders – we presume as a result of climate change but there is no real evidence for that.' Foremost among the suspects is the strikingly adorned Wasp Spider *Argiope bruennichi*, which, prior to 1990, was known only from a few sites along the south coast of England. This European species was first recorded in the UK in the 1920s but remained confined to the far south for perhaps 60 years. However, by 2016 it was widespread across a large swathe of southern and eastern England, as far north as Lincolnshire.

The mildly venomous Noble False Widow Spider *Steatoda nobilis* is another relatively recent arrival in the UK. It was originally native to Madeira and the Canary Islands but has spread extensively in Europe and

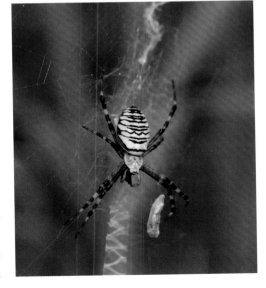

Wasp Spiders have been in the UK for a century, but only since the 1990s have they spread out from their longstanding south-coast stronghold.

elsewhere, presumably via accidental, human-mediated transport. First seen in the UK in the 19th century, it has made colossal gains in recent years, from just three recognised sites along the south coast in 1990 to multiple records across England as far north as Cumbria. Over the same period the Daddy-longlegs Spider *Pholcus phalangioides*, a regular dweller in human habitations, has infilled its range in England and is seen increasingly often in Scotland. Finally, a large and rather unwelcome (on account of its painful bite) tube web spider *Segestria florentina* has also increased substantially within southern England, again mostly by range infilling. It was initially reported in 1900, and recent increases in distribution and abundance of this European arachnid may well be a consequence of climatic amelioration.

How can spiders move so far, so quickly? One explanation is incidental transport with garden plants, but another, rather more romantic, one involves 'ballooning'. This remarkable phenomenon results in spiders casting their fate to the wind, not by being accidentally swept up in a passing gale but by a deliberate dispersal strategy. Individuals climb to as high a vantage point as they can find, maybe at the top of a plant, and, balancing on tiptoe, protrude their abdomens upwards. A silken thread is exuded to catch the wind, and the spider lifts off and rides the skyways until making landfall sometimes tens or even hundreds of kilometres from lift-off. Ballooning spiders have been caught at altitudes as high as 5km and sometimes drop in on ships far out at sea. It is hard to imagine a riskier way of getting about, and mortality rates must be enormous, but many species do this and evidently it works well enough. No wings needed, then, for potentially rapid range expansions.

For some groups of invertebrates, however, there is no way that aerial performances could influence range increases. The Conchological Society of Great Britain and Ireland has run mapping schemes for over a hundred years, generating a lot of data on the distribution of slugs and snails in the British Isles. Some changes in distribution are apparent, but Peter Topley of the Conchological Society advises caution in interpreting them, not least because of variations in recorder effort over time and space. Nevertheless, there are some interesting goings-on. The introduced Girdled Snail *Hygromia cinctella*, originally a Mediterranean species, appeared in Devon in the 1940s. Since the 1970s it has shown a marked distribution expansion to the east and north of the country, and by 2008 it had reached Glasgow. It is a 'garden' species and is probably spread primarily via human activity rather than as a consequence of improved climatic conditions. As a distraction from serious science we

Freshwater and terrestrial vertebrates

chapter four

Vertebrates are, in general, more familiar to most people than plants and invertebrates. In consequence, we know more about the ecology of these animals, and the prospects of identifying effects of climate change on them are higher than for most other fauna and flora. Even so, responses are likely to be varied because ectotherms (fish, amphibians and reptiles) face different challenges from those encountered by endotherms (birds and mammals). And not all vertebrates are well studied, though birds stand out as recipients of more attention than any other group of animals.

The UK's vertebrate fauna is small, almost trivial in number compared with plant or invertebrate diversity. Altogether, across freshwater and terrestrial habitats, there are about 400 species of fishes, amphibians, reptiles, mammals and breeding birds. Of this number, birds constitute more than half the total. There are broadly similar numbers of fishes and mammals (50–60 species of each) but a highly impoverished herpetofauna with a combined total of just over 20 amphibians and reptiles. These numbers include established, introduced species as well as natives and are necessarily vague because some categories, such as occasional bird visitors, are hard to delimit.

Introductions have occurred, and proved successful, in all five vertebrate groups, and a few, such as the Brown Hare *Lepus europaeus*, Rabbit *Oryctolagus cuniculus* and Fallow Deer *Dama dama*, are longstanding; the equivalent, perhaps, of plant archaeophytes. Most newcomers, though, would classify as neophytes, having arrived within the last couple of centuries. In this chapter I exclude marine

OPPOSITE PAGE:
The unmissable Great White Egret, a recent coloniser spreading rapidly across much of England.

mammals and seabirds, the latter constituting – again depending on definition – perhaps a quarter or more of the UK's bird fauna. All these species are considered in Chapter 7's deliberations on coastal and marine communities.

Although small in number, the vertebrates include many of the British animals that are best known to the general public. There are even popularity contests, one of which for birds in the UK was recently won by the Robin *Erithacus rubecula*. Popularity does not, however, always translate into ease of study or observation. Freshwater fishes are well known to the angling fraternity when hauled out of their habitat on the end of a line, but are mostly hard to spot in the water, let alone identify. For this reason they are perhaps the group least familiar to most people. Amphibians and mammals tend to be secretive, nocturnal or both, and reptiles, though diurnal, are easy to miss by an untrained eye. Birds are of course the exception and, largely due to the popularity of birdwatching, they account for by far the most extensive studies of all of the UK's vertebrates, indeed of all the country's fauna. They also add a unique aspect to wildlife observation because many of them have regular and impressive migratory habits, spending only limited amounts of time in the British Isles each year. Arrivals are eagerly awaited in spring and autumn and are therefore potential responders to climate change. An impressive database including many aspects of British bird ecology has been accrued by the British Trust for Ornithology (BTO), extending over many decades and constituting a valuable resource for investigating possible effects of climate change.

As ectotherms, with activity directly related to ambient temperatures, a reasonable supposition is that fishes, amphibians and reptiles should be more sensitive to climate change than the endothermic mammals and birds. This may be true, but even though few mammals and no birds hibernate, all are influenced in various ways by the severity of winter and the earliness, or otherwise, of spring. The situation for migrant birds is particularly complex because weather in their summer (winter visitors) or winter (summer visitors) quarters, as well as what they experience during migration, is likely to have effects that cannot be detected by measuring changes in the British climate alone. In many years this is demonstrably true of bird migration, where the animals wait for a favourable wind direction before making mass arrivals in the UK.

General phenology

Difficulties of observation are no doubt the reason that the *Nature's Calendar* website includes no fishes, reptiles or mammals among the species for which records are routinely collected (see overleaf). Some amphibians are included, but birds feature by far the most prominently, and within that group the main emphasis is on first arrival dates of migrants. This makes good sense, because other aspects of bird phenology such as first nest-building and feeding of young are more easily missed than the call of a Cuckoo. The long bird list capitalises on the abilities of a large cohort of naturalists skilled in their identification. Birdwatching is far and away the most popular type of participatory natural history in the UK, exemplified by the Royal Society for the Protection of Birds (RSPB) membership of more than a million people. This is the kind of number political parties can only dream of, and of which they might take a bit more notice on the rare occasions when they develop policies for environmental protection.

Spawning time for Common Frogs, an early spring affair to watch out for.

Vertebrate phenology measures in the UK

Group	Species	Spring/summer event	Autumn event
Amphibians	Common Frog *Rana temporaria*	First spawn First tadpoles	
	Smooth Newt *Lissotriton vulgaris*, Palmate Newt *L. helveticus* and Great Crested Newt *Triturus cristatus*	First observation of any species in ponds	
Birds	Blackbird *Turdus merula*	First nesting and feeding of young	
	Blackcap *Sylvia atricapilla*	First arrival (except where the species overwinters)	
	Blue Tit *Cyanistes caeruleus*	First nesting and feeding of young	
	Chiffchaff *Phylloscopus collybita*	First arrival	
	Cuckoo *Cuculus canorus*	First arrival	
	Fieldfare *Turdus pilaris*	Departure dates	First arrival
	Great Tit *Parus major*	First nesting and feeding of young	
	House Martin *Delichon urbicum*	First arrival	Departure date
	Nightingale *Luscinia megarhynchos*	First arrival	
	Redwing *Turdus iliacus*	Departure date	First arrival
	Robin *Erithacus rubecula*	First nesting and feeding of young	
	Sand Martin *Riparia riparia*	First arrival	Departure date
	Song Thrush *Turdus philomelos*	Winter singing	
	Spotted Flycatcher *Muscicapa striata*	First arrival	
	Swallow *Hirundo rustica*	First arrival	Departure date
	Swift *Apus apus*	First arrival	Departure date
	Turtle Dove *Streptopelia turtur*	First arrival	
	Wheatear *Oenanthe oenanthe*	First arrival	
	Whitethroat *Sylvia communis*	First arrival	
	Willow Warbler *Phylloscopus trochilus*	First arrival	

Source: the Woodland Trust's *Nature's Calendar* website.

Freshwater and terrestrial vertebrates

Blue Tit nesting has received a lot of attention, because climate change might cause chicks to hatch out of sync with their insect food supply.

General distribution and abundance

Vertebrate distributions are among the best documented of all the UK's fauna. There have been many changes in recent decades but most of these are unrelated to climate. Increases in birds of prey have usually reflected a combination of recovery from pesticide damage suffered in the mid-20th century and a reduction, sadly by no means universal, of persecution by gamekeepers. The recolonisation of England by Buzzards *Buteo buteo* is perhaps the most obvious example of this improvement in status, with the Red Kite *Milvus milvus* not far behind. Likewise the ongoing spread of Polecats *Mustela putorius* across much of England, from a relict stronghold in Wales, can be attributed to a reduction in persecution. On the other hand, declines in distribution or abundance, as always more common than successes, are usually due to habitat deterioration of one kind or another. The question is whether climate effects on the recent fate of any of our vertebrates can be discerned above a background of multifarious other causes of status change.

Freshwater fish

Phenology

Because these animals are so hard to see, there have been few opportunities to look for phenological changes in freshwater fish. An early plan when the British government became interested in climate change during the 1990s was to include, among other measures, recording the timing of spawning migrations by Atlantic Salmon *Salmo salar*. This seemed like a good idea, because salmon become conspicuous by prodigious leaping as they surmount weirs and waterfalls on their way upstream to headwater spawning redds. Despite the fact that numbers are counted annually on some rivers, such as on the Tummel at Pitlochry where they register on an automatic machine as they negotiate a fish ladder, the plan came to nothing. It turned out that the migration patterns were too complex, and recording not sufficiently reliable, for this feature to be used as a phenological indicator. Salmon migrate at different times in different rivers, often with discrete spring and autumn runs and something of a hiatus in between, although in Ireland most salmon run in summer. However, the most obvious trend in recent decades has been for spring migrations to decline rather than for the fish to change their phenology. Moreover, salmon behaviour in rivers is dictated more by immediately preceding rainfall spates rather than by temperature. These downpours vary unpredictably between years and vary locally across the country.

Failure of this apparently promising approach to looking at fish phenology left little in the way of viable alternatives. Many freshwater fishes spawn in spring, but their activity, deep among the fronds of undulating waterweeds, is rarely witnessed. Fortunately, rarely has not meant never. Windermere in Cumbria has a long history of ecological investigations, not least because the Freshwater Biological Association (FBA) has its main laboratory there. Probably no other lake in the world has attracted so much scientific attention for so long. Since 1929 the FBA has studied pretty much every conceivable aspect of Windermere's ecology, including its fish populations. One part of the FBA's programme has included estimating the spawning seasons of a common inhabitant, the Perch *Perca fluviatilis*. This dashing little predator is abundant in Windermere, and many are trapped throughout the year to measure various aspects of their population

ABOVE: Windermere, one of the best-studied lakes anywhere in the world.

LEFT: Perch have advanced their breeding season in Windermere as the water has warmed, but evidence from other water bodies, and for other fish species, is lacking.

dynamics. During the spring, live-trapped Perch are recorded as either being still with eggs or as having laid them, thus giving a quantitative estimate of the main spawning period. Since the mid-1940s peak spawning time has advanced by about two weeks, with the biggest changes happening in the 1970s and 1980s. In the early days most Perch fry hatched in early June, but by 2010 Ohlberger *et al.* (2014) found that they could be caught in the third week of May. This altered phenology coincided with a substantial warming trend of around 2°C in Windermere's surface water over the same period. No doubt this type of climate-related advance has been going on all over the country, and for other fish as well as Perch. The study amounts to a tantalising taster of what must have been happening on a large scale, an inference that unfortunately cannot be supported by more extensive evidence.

Distribution and abundance

Aside from garden and forestry plants, freshwater fishes are moved around by humans to a greater extent than any other wildlife in the UK. For decades, probably centuries, the angling fraternity has stocked ponds, lakes and rivers throughout the land with species they find desirable to pursue. Earlier still, freshwater fishes were an important part of the medieval diet and carp ponds were created to supply that need by monasteries and local estates. More recently there have been novel additions to our fish fauna, including the Zander *Sander lucioperca*, originally from mainland Europe and introduced by persons unknown. This enormous and largely undocumented flow of animals around the country dwarfs any natural range expansions that might have occurred.

There is also the inherent difficulty faced by all freshwater fishes excepting the few, mostly migratory, species that can survive in seawater. How, even under otherwise favourable conditions, could they move unaided between isolated ponds, lakes or rivers? Historically, most freshwater fishes in the British Isles probably attained their core distributions shortly after the Ice Age ended, exploiting the large expanses of freshwater and interlinked river systems that became widespread at that time, but were short-lived as the glaciers melted and sea levels rose. Since then there can have been few opportunities for natural dispersal, and one much-vaunted idea that fish eggs might stick to the feet or feathers of wildfowl has little empirical support despite having been proposed regularly for a very long time. Personal experiences with garden ponds have reinforced my cynicism about this dispersal mechanism. For more than 20 years a population of Three-spined Sticklebacks *Gasterosteus aculeatus* remained confined to a single pond in my garden, within close proximity of four others. This was despite regular use of all the ponds for drinking and bathing by a profusion of garden birds, admittedly not including wildfowl but nevertheless all with feet and feathers.

For most fish confined to ponds and lakes, 'natural' dispersal must be a very rare event. There is, however, an option for stream- and river-dwelling species to change their preferred altitude. This type of change should be relatively insensitive to direct human interference and has been observed in both the UK and France. In a wide-ranging study comparing the distributions of 32 stream and river fishes across the whole of France, between sampling periods 1980–1992 and 2003–2009, big changes were found by Comte & Grenouillet (2013). Most of the fishes they studied also occur commonly in the UK. In total 25 species moved upstream, and range-centre shifts per decade, averaged across all species, were to almost 14m higher altitudes and to 0.6km further upstream, compared with the situation that previously pertained. Among those moving up and along were Atlantic Salmon, Grayling *Thymallus thymallus*, Silver Bream *Blicca bjoerkna*, Ruffe *Gymnocephalus cernua*, Tench *Tinca tinca* and Ten-spined Sticklebacks *Pungitius pungitius*. Grayling, Salmon and Burbot *Lota lota*, as well as some other species, experienced overall range contractions. The changes in range centre were mostly due to range extensions at high altitude rather than contractions in lower reaches, a difference comparable with that seen in butterflies between northern and southern range margins. Dispersal at the upper range limits extended by an average of more than 60m of altitude per decade, whereas range contractions at lower limits were some tenfold slower. Water temperatures rose commensurate with, but faster than, the changes in range centres. Although most species extended their ranges, some 30 per cent of typically 'high-elevation' fishes, mostly salmonids, did the opposite. Evidently cold-water species were suffering from the warming environment, and in general fish responses throughout the river systems were not keeping up with climate change, potentially boding ill for the future.

Although major range extensions in freshwater fishes are unlikely and difficult to assess, changes in abundance might be more accessible. A theoretical expectation is that species thriving in the relatively sultry waters of lowland England, such as members of the carp family, should do well in a warming climate or at least not be adversely affected by it. So far there is no tangible evidence of any effects, but then again precious few fish populations are monitored sufficiently well to detect increases over time. The regular movements and restocking of many species that go on year on year would make natural changes hard to discern.

More worrying, although still mostly a theoretical concern in the UK, is the risk of population declines among those fishes that thrive in cold lakes or river headwaters. Warming trends comparable with those in Windermere have been seen in all freshwater habitats where measurements have been made and, as discussed in Chapter 3, are already disadvantaging invertebrates such as mayflies upon which many fishes feed. Salmonids are among the most widespread cold-water

Brown Trout are cold-water fish at long-term risk from climate change but they may outcompete less heat-tolerant species in the short term.

species, particularly Brown Trout *Salmo trutta* and Atlantic Salmon. Declines of both these magnificent fishes in the River Wye have been attributed, at least in part, to increases in water temperature of between 0.5 and 1°C between 1985 and 2004. Cool, well-oxygenated waters are vital to the survival of most life stages of trout and salmon, but many other factors, including pollution and predation pressure, also affect their population dynamics. In the case of salmon, events at sea may be particularly important, and these are considered more thoroughly in Chapter 7. The UK is situated about halfway between the southern and northern range limits of salmon rivers on the European side of the Atlantic, and these fish still breed successfully as far south as they have ever done within historical times, in northern Spain and Portugal. Presumably it is at these southern outposts that any damaging effects of climate warming will show up first. For now, direct impacts of warming freshwater habitats on British salmonids remain unproven, but because of the importance attached to the future of these animals, concerns about their future have become the subject of much speculation in computer models, as discussed in Chapter 9.

Just one freshwater fish, the Burbot, has gone extinct in the UK in recent times, and there have been suggestions that climate warming contributed to its demise. The arguments behind this suggestion are unconvincing. Burbot are cool- but not cold-water fish, and until the mid-20th century were reasonably common in lowland rivers of eastern

Arctic Charr, an uncommon cold-water fish definitely in decline.

England, particularly in the Fen country. The last definite sighting was in the late 1960s, before the main period of climate change that came decades later. Moreover, the Burbot's home rivers were subject to many serious pollution events and major engineering projects that altered channel sizes and water flows. These machinations seem much more probable causes of the Burbot's demise than anything to do with the weather.

The most likely future fish victims of climate change in the UK are those cold-water species confined to deep mountain lakes. The Arctic Charr *Salvelinus alpinus*, once common enough to support a commercial fishery, has declined in most of its haunts as temperatures rise. Two rare species of whitefish are confined to a very few lakes in the UK and look to be at particular risk. Whitefish are distant relatives of trout and salmon and, like them, thrive best in cold-water systems. The Powan *Coregonus lavaretus* survives in northern England, Scotland and Wales but only in a few lakes in each country. Rarer still is the Vendace *Coregonus albula*, which now persists only in Bassenthwaite Lake (just about) and Derwentwater in Cumbria. Undoubtedly the UK's rarest fish, within the last century Vendace disappeared from their two only other recorded locations, both in Scotland. All our whitefish face problems of pollution and dangers from introduced predators such as Pike *Esox lucius*, and warming water is the last thing they need if they are to remain part of the British fauna.

Amphibians

Phenology

The UK's impoverished amphibian fauna of just seven native species has punched above its weight with respect to identifiable impacts of climate change. All of them have either made phenological compensations or suffered damages to their population dynamics since the arrival of warmer and wetter winters that escalated in the 1980s. Even two of the three most successful non-native species in the UK have responded along similar lines.

All British amphibians hibernate during the winter months, usually in refugia somewhere on land, and migrate to their breeding ponds in spring. Spawn is deposited in the water soon after the animals' arrival, and adults then return to their terrestrial habitat within a few weeks of egg-laying. Two signals of critical phenological events are therefore migration and spawning. Migration occurs mostly at night and is usually hard to quantify, but a series of garden ponds in my Brighton garden provided ideal circumstances for detection of animals as they moved in. Only having to walk out of the back door allowed inspection of the ponds, with a powerful torch, almost every night in winter and early spring from 1978 to 1994. First arrivals of individual Smooth Newts, Palmate Newts, Great Crested Newts and the non-native Alpine Newt *Ichthyosaura alpestris* (abundant in these ponds) were noted, not initially with any expectation of seeing changed phenology. All four species, however, showed dramatic shifts to earliness over the 17-year period, such that by 1994 the first arrivals were coming in 5–7 weeks sooner than they did in the 1970s. From a typical early spring event with newts turning up from mid-February onwards, migration of vanguard individuals became, by the early 1990s, a mid-winter one. By then it was normal to have newts in the garden ponds by Christmas Day. The three native species responded more or less in parallel, while the Alpine Newts showed the same pattern of increasing earliness but with a delay, maintained over the years, of about four weeks relative to the other three newts.

By contrast, Common Frogs which also used the ponds showed no significant change in spawning time over the same period, and Common Toads *Bufo bufo* in Dorset also showed no trend towards earliness in the late 20th century. Non-native Pool Frogs *Pelophylax lessonae* bred in the same garden ponds, but these animals require much warmer

Garden ponds, such as this one of mine in Sussex, are convenient facilities for monitoring changes in amphibian phenology.

conditions for breeding than the native species. In the late 1970s Pool Frog spawn was typically laid around mid- to late May; by 1994 it was often appearing by mid-April. In Hampshire, Natterjack Toads were also breeding earlier by 1994 than they had in the late 1970s, although the change was again less marked than that noted with newt migration times, as shown by Beebee (1995). The Natterjack population was intensively monitored as an ongoing research project, during which first spawn sightings shifted, on average, from late to early April.

In all these examples, trends correlated with biologically relevant temperature changes at local recording stations. In the case of Smooth Newts, for example, first arrivals correlated with maximum temperatures in the previous month. These temperatures increased at a rate of around 0.24°C per year, so by no less than 4°C overall. Perhaps surprisingly, rainfall in any proximal month had no detectable effect on migration times.

Caveats on interpreting these changes in terms of climate effects are at least twofold. Firstly, the observations relate to just three locations in the south of England, and secondly, in the case of newt migrations only the first arrivals rather than the bulk of the populations were recorded. Subsequently, more detailed investigations into the migration of Smooth and Palmate Newts to a pond in mid-Wales, using pitfall trapping, quantified the mean arrival times of the entire populations.

These had also advanced, but only by around 1–3 weeks between the 1980s and the late 1990s, so by substantially less than my garden first-comers. Elsewhere, mostly anecdotal reports have confirmed that newts are migrating earlier than was previously normal all around the country. It looks like a real climate effect, and for newts in the garden ponds the trend continued, albeit more slowly, into the early 2000s.

Frogs turned out to be more interesting than the garden suggested. The Royal Meteorological Society, in a far-sighted move, began to collect the dates of natural events including first frog spawn sightings as long ago as 1926. Maxwell Savage, one of Britain's pioneering herpetologists, compiled all the records between 1938 and 1947 to create a UK distribution map of spawn dates, with contours separating districts according to their earliness. This showed that eggs appeared first in the far south-west, often in January in parts of Devon, and became progressively later going north or east. The very latest spawning was around mid-April at high altitudes in the Lake District. More recently, data from the UKPN records of spawning times between 1998 and 2007 were investigated and compared with Savage's map of more than half a century earlier, by Carroll *et al.* (2009). This revealed that at the national level Common Frogs have indeed started spawning earlier, requiring a redrawing of Savage's contour map and showing an average advance of about ten days since the mid-20th century. Furthermore, this change correlated with mean January–March temperatures and indicated an advance in spawn dates of about five days for every 1°C rise in temperature. This is a good example of how looking at large, geographically dispersed data sets can demonstrate effects that may be equivocal if just judged locally or within short timescales. Common Frog spawn is now regularly found before Christmas in parts of south-west England.

Distribution and abundance

There is little scope for range extension for most of our native amphibians because all three newts, Common Frogs and Common Toads already extend across the whole of the UK, from northern Scotland to the south coast of England. Natterjack Toads have a much more restricted distribution because they are confined to the specialised habitats of lowland heaths or, more commonly, coastal sand dunes and upper saltmarshes. This essentially precludes range expansion, because these habitats are invariably isolated by the sea or by tracts of unsuitable terrain, mostly farmland. Unless this habitat specialisation is relaxed, which seems unlikely in the foreseeable future, Natterjacks are unlikely to benefit from climate change. They may, on the contrary, suffer consequential population declines. Mild, wet winters increase toadlet mortality during hibernation, though it is not yet clear whether this threat will prove important in the long term. More ominously, sea-level rise has increased the frequency of high tides in saltmarsh habitats of north-west England and on the Scottish Solway, squeezing the toads into an ever narrowing strip of marshland, inland of which are unsuitable agricultural fields. Scottish Natterjacks in particular have declined recently, and this may be due at least partly to ever more tidal inundations of their key habitat.

It is not just Natterjack Toads that are suffering from milder and wetter winters. Chris Reading (2007) followed the fate of a population of Common Toads at a site in Dorset between 1983 and 2005, noting that body condition of females after hibernation was reduced by warm winter weather. This was paralleled by trends over time of reduced fecundity and increased mortality, all concomitant with a continuous population decline. Similarly, Richard Griffiths and his colleagues at the University of Kent related reduced winter survival of adult Great Crested Newts to warm and wet winters in a metapopulation (meaning a population using several different ponds) undergoing overall decline. Recruitment into the newt population was sporadic and mostly focused at one of the multiple breeding ponds.

Great Crested Newts can experience high mortality in wet and mild winters.

This investigation by Griffiths *et al.* (2010) highlighted the importance of long-term adult survival, allowing multiple breeding opportunities, if the metapopulation is to persist. Both of these studies relate to single, confined areas, and it remains to be seen as to how general these effects are. Suffice it to say that both Common Toads and Great Crested Newts have been declining nationally in recent decades more than the other widespread amphibian species.

Only one species, or more accurately 'species complex', of amphibian in the UK seems to be improving its lot as summers warm up. European Pool Frogs and Edible Frogs *Pelophylax esculentus* are close relatives, with the Edible Frog constituting a hybrid between Pool Frogs and Marsh Frogs *Pelophylax ridibundus* that, due to a highly unusual genetic system, can persist with either one of its parent species but does not need both. Such a mixture of Pool and Edible Frogs was introduced to a set of several neighbouring ponds in Surrey at the start of the 20th century and remained confined to that area for decades thereafter. In the early 1990s a change was afoot. The frogs started appearing in new places, often linked by ditches or rivers to the original site, and have continued a slow hop across central Surrey ever since. This is one of the few cases where it is possible to discern a causative link between changed phenology and abundance. These frogs have a much longer tadpole phase than our native species and, until recently, many failed to metamorphose before autumn in an 'average' year. Those larvae that remained in the ponds generally died, meaning that in many years there was little or no breeding success. Earlier

Edible Frogs are non-native amphibians that have advanced their spawning time and widened their distribution in recent decades.

spawning times, and probably warmer water, have transformed this situation, and most years now see good recruitment of young frogs to the population. Climate change is therefore promoting the spread of non-native amphibians, fortunately ones that have no known impact on other fauna.

Reptiles

Phenology

Given that they are sun worshippers, we might reasonably expect reptiles to benefit from a warming climate. Perhaps some do, but evidence to support that prediction is as yet scant indeed. All six of the UK's native species hibernate in frost-free refugia underground, and as the days lengthen they emerge bleary-eyed to enjoy early spring sunshine. At this time of year basking lizards and snakes are often conspicuous and would seem to offer ideal opportunities for phenology watchers. Alas, it is not easy to measure emergence times accurately because reptile spotting is a subtle skill that takes much practice to acquire. Even experienced reptile enthusiasts, of which there are rather few, sometimes have difficulty predicting exactly which weather conditions will provide good hunting opportunities. The upshot of this is that there are no systematically collected data sets about emergence from hibernation for any British species in the wild. All of them mate shortly after emergence but, unlike amphibians, this does not involve congregation at specific places, and breeding activity is even less amenable to regular observation than the end of hibernation.

We are not, however, totally bereft of phenological information about British reptiles, although everything we do have relates to just one species, the Sand Lizard, and most of that comes from studies in Sweden rather than in the UK. Sweden, like England, is at the northern range edge of this enchanting animal, and it seems reasonable to suppose that observations in Scandinavia are relevant to what happens here. By late April in both countries, vividly green-flanked male lizards court and mate with, and then attempt to mate-guard, the more dourly coloured females. Guarding is rarely very successful, and both sexes frequently copulate with multiple partners. In late May or early June females lay a clutch of eggs buried in sand, and a second clutch can follow a few weeks later. Long-term (9–15-year) studies in Sweden have shown that above-average spring temperatures correlate with early egg-laying, which in turn results in higher offspring fitness and subsequent survival. Not only that, but warm springs promote high levels of multiple paternity in Sand Lizards, and this also increases the proportion of viable juvenile lizards. It therefore seems likely that long-term warming trends may, by advancing these phenological traits, impact positively on Sand Lizard abundance.

A few observations in England, casual by comparison with the Swedish studies, support the idea that climate change is altering Sand Lizard phenology. For many years I maintained a small population of Sand Lizards as part of a captive breeding programme for conservation in an outdoor vivarium in the garden. This provided the same advantages for observation as did garden ponds for amphibians, since it was possible to look daily for the first lizards, invariably males, to emerge from hibernation. I recorded these appearances from 1978 to 1990, and over that period there was a statistically significant trend towards earliness. In the late 1970s I would see the first lizard sometime in mid-March, whereas by the late 1980s the animals were appearing by mid- to late February. And something else was going on. In the early years it was rare for a female to lay two clutches of eggs, but by the 2000s it had become normal practice. Was this a feature of captivity, or were events in the vivarium mirrored in the outside world? Longstanding Sand Lizard experts Keith Corbett and Jon Webster, of Amphibian and Reptile Conservation, think that similar changes were afoot in wild populations. In Jon Webster's words:

> *I can report from my diaries that during the 1970s there is no reference to finding juvenile lizards before September, in fact the first diary entry of note is 5th August 1984 when Mike Preston and I found juvenile Sand Lizards on Gong Hill, and Bill Whitaker had found juveniles in Dorset the week before. Another interesting entry is finding a female Sand Lizard in Dorset on a sunny day in February 1989 – that is the earliest record from my diaries, as before emergence was recorded from early March onwards. Generally egg-laying occurs at the very end of May and more usually the first week in June, but I have another entry for 1990 where apparently Mike Preston dug up a clutch of eggs at Churt on or around 12th May, and I have seen female lizards on Crooksbury a week later that have clearly laid their eggs. I clearly remember noticing juveniles of different sizes and remarking that these must be as a result of double clutching, I think this was around the 1990s.*

Freshwater and terrestrial vertebrates

Although unsupported by systematic records, both of these observers concur that double clutching by wild lizards was virtually unknown in the 1970s but had become commonplace by the 1990s and 2000s. So although we have seriously imperfect information about reptile phenology in the UK, what little we do possess indicates that changes have been under way in recent decades.

Distribution and abundance

Evidence for climate-induced population or distribution changes among British reptiles is hardly any better than that for changes in phenology. The Viviparous Lizard *Zootoca vivipara*, the Slow-worm *Anguis fragilis* and the Adder *Vipera berus* have always ranged the full length of Britain from southern England to northern Scotland. The lizard and the snake are the most cold-adapted reptiles in the world and both can be found well north of the Arctic Circle in Scandinavia. It is notable, therefore, that the Adder has declined in the UK more than any other reptile, and at least in some parts of the country the Viviparous Lizard also seems to be in trouble. Southern biogeographical range limits for both of these species, aside from some high-altitude populations, are in northern France not very far from the English Channel coastline. Could climate

Viviparous Lizards are not far from their southern biogeographical range margin in England and might be disadvantaged by a warmer climate.

warming be driving declines in these cases? It might have a role to play but, at least in the case of the Adder, probably not a prominent one. This strikingly marked snake is suffering, especially in central England, from habitat deterioration and population isolation, factors almost certainly of greater importance than warmer weather. The Viviparous Lizard is perhaps a more likely victim of extra heat. Laboratory studies have shown that an average temperature increase of 2°C above current levels, which is one prediction of continued, rapid climate change, dramatically increases adult mortality. Also relevant is that long-term monitoring of reptile status in the Netherlands has revealed that, since the mid-1990s, Viviparous Lizards have declined substantially and much more than any other reptile in that country. The fact remains, though, that we do not really know the reason for Viviparous Lizard declines and they are certainly not happening everywhere. In the French Pyrenees, body size and survival rates of this charming little lizard were on the up during an 18-year study that coincided with a period of increasing May temperatures.

Two other British reptiles, the Sand Lizard and the Smooth Snake *Coronella austriaca*, are habitat specialists that occur on lowland heaths and, more rarely, on coastal sand dunes. Both have very limited distributions but, at the northern edges of their biogeographical ranges in the UK, both could potentially increase in numbers and broaden their distributions in a warmer climate. That Sand Lizards can survive and breed much further north than the limits of their current distribution is certain, because animals translocated to sand dunes on the Hebridean island of Coll in the early 1970s thrive there to this day. Population sizes of both species are, as far as we know, holding up well – and at least in the case of Sand Lizards could be benefiting from the changes in phenology outlined above, but we have no detailed information about that. Nevertheless, there are some clues. The Netherlands monitoring study showed that Sand Lizards, confined to habitats similar to those used in the UK, have increased consistently since the 1990s. There is even some preliminary indication that, in Dorset, the lizards are broadening their habitat niche around heathland edges. Recent surveys have found Sand Lizards in unusual locations. New populations are cropping up in habitats that would not traditionally suit them, such as development sites, railway lines, derelict brownfield sites and roadside verges. This spread may reflect warmer summers allowing more successful breeding in places outside their mainstream heathland habitats. Reptile expert Chris Gleed-Owen has found that:

Sand Lizards are increasingly turning up in the most unexpected places, including rubble piles and weedy building sites. It's no laughing matter for a developer, as work has to stop until a licence is obtained and the lizards are re-homed. These new locations tend to be contiguous with heathland or other known habitats, and they are often linear sites acting as colonisation corridors. This is great news for Sand Lizards because it reverses some of the fragmentation that has been isolating populations for decades.

Maybe there won't be too many tears shed about problems for developers. However, there is a caveat about what the observations really mean. Maybe the 'new' sites are places where nobody has looked until recently, and it is possible that the lizards were there before climate change kicked in.

The UK's sixth native species, the Grass Snake *Natrix helvetica*, is widespread and common in England and Wales but reaches its northern limit close to the border with Scotland. Here is a potential range expander, and Grass Snakes have been found recently in Dumfries and Galloway. However, critical investigation of verified historical records by Cathrine (2014) suggested that they may always have been there, but few in number and under-recorded. It seems that nothing dramatic has happened yet with Grass Snakes in Scotland.

Marginal habitat outside heathland that is now the residence of Sand Lizards in Dorset.

Introduced reptiles might also be expected to improve their lot in the UK as the climate warms, but so far only one non-native species, the European Wall Lizard *Podarcis muralis*, has established itself widely in Britain. There are now numerous, isolated but thriving populations of this energetic little climber dotted across southern and eastern England. Many are recently established and all must have been started by deliberate releases. No doubt the lizards are benefiting from warmer summers, but their chequered history makes it impossible to discern the extent to which climate is influencing their population dynamics. Some colonies have certainly been here for many decades before any recent warming trends became apparent.

Mammals

Phenology

As endotherms with control over their body temperatures, mammals differ strikingly from the terrestrial vertebrates discussed previously because most British species do not hibernate. Some do, including all the bats, Hedgehogs *Erinaceus europaeus* and dormice, but the rest get by with, at most, somewhat reduced activity in the winter months. End-of-winter events are therefore not sufficiently well defined, in most species, for meaningful phenology. It was realised as long ago as the 1970s that spring emergence times of Hedgehogs varied across the country in a way similar to frog spawning, being earliest in south-west England and later in Scotland, but only very recently have attempts begun to record the earliest Hedgehog sightings systematically. Unfortunately, discerning any trends may now be compromised by the rapidly diminishing Hedgehog population. Falling numbers reduce the probability of detecting the first animals to emerge, year on year, and this may mask any real shifts in phenology. Even so, Hedgehogs are popular animals and hopefully 'citizen science' will produce useful information about any impact of climate on them in future years.

Bats present a different problem. Many are easy enough to see as they flutter around our houses from spring through to autumn, but difficult to identify without proper training and bat recording devices. Eighteen species are now recognised as inhabitants of the UK, including several only discovered within the last 20 years, and all but one of them are regular breeders. The identification problem means that phenology of particular species, which is what we need to know, is not readily accessible to most naturalists. Information is therefore sparse, but not completely lacking – though what there is relates to summer breeding activity rather than spring emergence. A study of Greater Horseshoe Bats *Rhinolophus ferrumequinum* by Roger Ransome at Woodchester Mansion in Gloucestershire from the late 1960s to 2012 found that birth timing became significantly earlier over that period (Jones *et al.* 2015). Around 1970, young bats were appearing in

Greater Horseshoe Bats have advanced their timing of giving birth as the climate warms.

mid-July whereas by the 2000s the average birth date was about ten days before then. Elsewhere, a similar investigation, but this time of Daubenton's Bat *Myotis daubentonii* (also a British species) at a site in the Czech Republic between 1970 and 2012, yielded remarkably similar results. Observations of the first young bats ended up 12 days earlier at the end compared with the start of the study, and the timing related to increases in April temperatures. It seems very possible that many British bat populations are responding in a similar way, but we just don't know.

Summer events are the only reliable measures of phenology that we have for other mammals, and these are based on studies of two ungulates, Chillingham Cattle *Bos taurus* living under semi-wild conditions in a northern English park, and Red Deer *Cervus elaphus* on the Hebridean island of Rum. The white cattle have been at Chillingham since at least the mid-17th century and receive minimal interference from humans. They calve all year round, and details of their population dynamics have been recorded for some 60 years. During this period the median birth date of the herd advanced by about one day per year. This corresponded with a decline in summer calving (from 44 per cent to 20 per cent of the total) and a corresponding increase in winter births, from 12 per cent to 30 per cent. The increase in winter birthing correlated with early onset of grass growth the previous spring, which is when conception occurs. Unfortunately, calves born in winter survive less well than those arriving in summer, indicating a potentially serious negative long-term effect of climate change on this fascinating historical population of feral cattle (Burthe *et al.* 2011).

Chillingham Cattle have changed their average calving time in response to climate change.

The breeding cycle of Red Deer on Rum has responded to climate change in various ways, so far without any consequences for population dynamics.

The Red Deer of Rum must be the most intensively studied population of mammals anywhere in the world. Tim Clutton-Brock and his colleagues have examined virtually all aspects of the biology of these deer for more than 40 years. Phenological measurements over 28 years were extraordinarily thorough and included six reproduction-related traits: sexual receptivity and calf birthing dates in females, antler cast and antler 'clean' (when surrounding 'velvet' growth tissue is discarded), and rut start and rut end dates in males, as described by Moyes *et al.* (2011). All of these traits advanced by between 5 and 12 days, and were largely explained by changes in 'growing degree days' (the number of days in a season warm enough to support substantial plant growth), promoting increased proliferation of vegetation in spring and summer. Male antler mass increased over the 28 years, while the length of the rut declined because end dates advanced more than start dates, suggesting increased competition and perhaps more stress among male deer. Despite these changes, there were no corresponding improvements in offspring birth weight or survival, and nor were there any changes in average male breeding success. This was considered a possible scenario if rutting behaviour became asynchronous with oestrus. It seemed that although climate change precipitated substantive advances in several aspects of the phenology of Rum's Red Deer, these alterations were neutral with respect to population size and viability. Later studies, however, concluded that the lack of fitness changes was probably due to complex, antagonistic effects of climatic factors and population density, and that it would be wrong to assume the population was inherently insensitive to climate change.

Distribution and abundance

Information about climate effects on the distribution and abundance of British mammals is even scarcer than that on phenology. Perhaps this means that most species have not been impacted yet, or, like Red Deer, are relatively resilient to it – but a more likely explanation is that not enough work has been done on the subject. An interesting exception concerns Badgers *Meles meles* in Wytham Woods (Macdonald & Newman 2002). Between 1987 and 1996, Badger numbers in this famous Oxford study site increased from 60 to 228 and, as the researchers pointed out, this could not have been due to release from persecution since the wood has always been strictly protected, nor to habitat change. What might have underpinned the increase was a rise in Badger body weight in January that correlated with January temperatures. Certainly mild, wet winters make earthworms more accessible than cold frosty ones, and worms form the main component of Badger diet. Presumably heavier Badgers mean fitter Badgers, maybe increasing reproductive success. Badger distribution has extended northwards in Finland over recent decades, probably a consequence of warmer winters, and it may be that climate change has been part of the explanation for Badger population increases across the UK. Of course there must be other factors at play, but those concerned about rising Badger numbers, mostly farmers and scientifically illiterate politicians, might usefully add this factor into their equations.

For obvious reasons it is not possible for most terrestrial mammals to extend their ranges from neighbouring Europe into the British Isles. Bats, however, are an exception, and individuals of several species cross the English Channel occasionally. Of particular interest is Nathusius' Pipistrelle *Pipistrellus nathusii*, a migratory bat recorded historically as an occasional visitor to the UK. Recently there have been increasing records of it in southern and eastern England, particularly in spring and autumn, and the discoveries of maternity roosts in England and Ireland have confirmed this bat as a resident breeder. Its recorded localities in the British Isles are experiencing increasing minimum temperatures, reduced inter-seasonal temperature variation and intermediate rainfall, and sites like this have multiplied since 1940, as recognised by Lundy *et al.* (2010). This expansion of suitable conditions is predicted to continue, and Nathusius' Pipestrelle may well be an example of a mammalian colonist benefiting from improved climatic circumstances in the UK.

Birds

Phenology

Some three million people in Great Britain engage in birdwatching every year, and it is no surprise that birds are by far the best known of the country's vertebrates. The RSPB's 'Big Garden Birdwatch', for which contributors need not even go outdoors, typically records several million bird sightings every January. Together with the observations of professional researchers, the vast mass of data now available about British birds has allowed the BTO and other organisations to investigate many aspects of how these animals are faring.

Two aspects of bird phenology have attracted attention. Nest-building together with egg-laying is, in many species, readily detectable as adults pick up material and fly with it to the construction site. Many birds breed in garden hedges, trees and shrubs so, rather like recording frog spawn in garden ponds, it is an indicator of spring that is accessible with minimal effort and reasonable precision. But arrival and occasionally departure times of migrant birds are also very popular as phenological indicators. The UK delights in two quite separate rounds of bird migration. In spring, species intending to breed here arrive from warmer climes, including several long-distance travellers that overwinter somewhere in Africa. Their departure in autumn is matched by invasions of birds that breed in the Arctic and descend on the British Isles to benefit from our mild winters, after which they head north again in spring. Just as colonisation is easier to report accurately than final extinction, so it is that noting arrival dates is likely to be more exact than observing departures, resulting in greater emphasis on the former measure. There is also a complication with interpreting migration records in the context of British climate, because arrival times might be influenced by conditions far from the British Isles. It is well known that unfavourable wind directions can delay the appearance of migrating birds by days or even weeks. At the very least, such local weather patterns will generate background noise with respect to detecting any trend for migration patterns to change consistently over time.

What, then, has happened to bird breeding cycles? Advances in the timing of egg-laying were among the first recognised signs of climate change's impact on British wildlife, as shown by Crick & Sparks (1999). Most (53 per cent) of the 36 species for which data were available over the 57 years from 1939 to 1995 advanced their median laying dates (MLDs), and of these 19 birds, 17 were clearly influenced by weather, particularly temperatures in March and April. Starting so long ago, it was interesting to realise that in the first decades of recording, corresponding to a brief climatic cooling, breeding became later but recovered in the 1980s and early 1990s to become substantially earlier than it had been in the 1930s. These trends have continued across the UK (see figure below). A comparison of 25 birds in Scotland between 1966 and 2003 revealed significant trends towards earlier egg-laying for five of them, namely Greenfinch *Chloris chloris*, Great Tit, Oystercatcher *Haematopus ostralegus*, Redstart *Phoenicurus phoenicurus* and Dipper *Cinclus cinclus*, while a further 13 trended non-significantly in the same direction. Egg-laying time for four of the above species (not the Dipper) correlated with mean monthly temperatures in one or more of the preceding three months. The average advance of all the study species between 1966 and 2003 was about four days.

The Redstart is one of several birds in Scotland breeding earlier than was normal in the past.

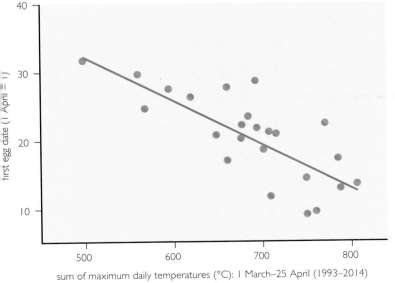

Egg-laying by Great Tits in Cambridgeshire advanced between 1993 and 2014 as spring temperature increased. The sum of maximum daily temperatures across the egg-laying period reflected the overall warming trend. Adapted from Hinsley et al. (2016).

These changes have been noticed by seasoned naturalists as well as professional scientists. In Chris Packham's words:

> *For a period in the early seventies* The Observer's Book of Birds' Eggs *was my spring bible – always in my pocket, very well thumbed as I searched suburban Southampton and the nearby countryside for nests. There were approximate laying times for each species: early/mid/late April etc. and they were accurate in terms of the appearance of their first clutches. But now they are pretty much all 'wrong' with most laying at least half a month earlier, some a full month. So in the time I've been interested in birds they've radically and significantly modified their behaviour – and my little brown manual is defunct!*

It is probably fair to say that most British birds have advanced their breeding seasons over the past few decades. The Robin, one of the species recorded for *Nature's Calendar*, had already started egg-laying an average of 15 days earlier by the mid-1990s compared with the 1940s.

For some of the changes, particularly for Blue Tits and Great Tits, there has been much interest in whether problems have arisen with food supplies. Is chick hatching still synchronous with caterpillar abundance? This is one of many potential disturbances that climate change might create at community levels, the subject of Chapters 6 and 7. For migrants, the situation is inevitably more complicated. A large body of research is now available to illuminate how climate change is affecting bird migration, breeding cycles and population dynamics, all brought together by Pearce-Higgins & Green (2014). Passerines have experienced the biggest changes in their MLDs. In the case of migrants such as the Pied Flycatcher *Ficedula hypoleuca*, however, the situation is complex because MLD depends not just on temperature at the breeding site but also on arrival time, and thus on conditions during migration. Consequently, for medium- and long-distance migrants there is a much weaker link between MLD and breeding-site temperature than is the case for short-distance migrants. In the UK, MLDs of Pied Flycatchers advanced by about nine days between 1974 and 2001, from around mid-May to the first week of that month.

Arrivals of spring migrants have generally become earlier in recent decades. Because birds are well monitored it is often possible to distinguish first arrival dates (FADs) from median arrival dates (MADs), the times by which more representative samples of the populations

Pied Flycatchers are among summer migrants that both arrive sooner and breed earlier than they used to do.

are present in the UK. Averaged across multiple species, FAD has advanced by about 0.24 days per year and MAD by 0.15 days per year. This kind of difference between FAD and MAD is similar to that seen for newts arriving in ponds (see Amphibians section on p. 130), confirming that some individuals will always be ahead of the game. Not surprisingly, the MADs cover a wide range of variation among species, and those travelling the greatest distances, such as Swallows from South Africa, have responded less strongly to climate change than those starting off closer to the UK, such as Chiffchaffs coming from central Europe and around the Mediterranean. Nevertheless, long-distance migrants are also arriving earlier than they used to do. First sightings of Swallows alighting in England were reasonably consistent, typically in the first week of April, between 1960 and 1990. After that they became consistently earlier, usually appearing in the last week of March by the early 2000s, a trend that mirrored average UK temperature changes over the same period. Swallow egg-laying dates followed the same pattern, becoming earlier by about ten days. For many short-distance migrants, advances in MLD have been greater than advances in MAD, thus shortening the lag between the two dates. Arrival times of short- and medium-distance migrants, starting their journeys in Europe or north Africa, relate to winter mildness, whereas the MADs of birds coming from further afield are influenced by more distant conditions, including the weather in their overwintering grounds.

Climatic influences on autumn migration times are less clear-cut. This is probably because departure is dictated by factors other than climate, not least of which are the success and timing of breeding while the birds are with us. In years of low reproductive output, some birds tend to depart earlier than they otherwise would. This also applies to species not actually breeding in the UK, but that stop over in mid-migration. Arctic-breeding Curlew Sandpipers *Calidris ferruginea*, which call in en route between their summer and winter headquarters, are an example. On the other hand, summer visitors that arrive earlier in spring than they used to do may stay longer to rear more broods, and delay departure for that reason. There are differences between short- and long-distance migrants in this regard. Some short-distance species are remaining closer to their breeding sites as winters become milder and therefore safer. Blackcaps are now often seen in the UK throughout the winter. In my part of Somerset, winter is the only time we see them – though in this case it is probably not because of overstay but because they move in from Europe to escape continental conditions in the bleaker months of the year.

Curlew Sandpipers visit the UK when travelling between winter and summer quarters. The timing of their autumn migration depends on breeding success in the Arctic.

Cetti's Warblers are another remarkable success story, spreading rapidly since their arrival in the 1970s.

Not all newcomers are herons. Cetti's Warbler was first established in the UK in 1973, and extended its range spectacularly after that in southern and eastern England, reaching as far as Northumberland in 2011. Though not an easy bird to see, there is little difficulty in detecting its presence thanks to the male's astonishingly loud song, delivered in intermittent and unmistakable bursts.

A warming climate has generally been good news for the populations of most of the UK's resident birds. Long-term studies have investigated relationships between climate and key population parameters, especially survival and reproduction rates. Winter temperatures are the most important influence on survival of adults and juveniles, particularly of land rather than sea birds. Cold winters increase mortality both directly, due to exposure, and indirectly by reducing food supply. This is especially important for insectivorous birds, since invertebrates become much less active at low temperatures. A major disadvantage of endothermy is the need to maintain a high

Wrens suffer high mortality in cold weather, and their population has increased considerably as winters have become milder.

food intake to support metabolic heat generation, a problem many other vertebrates avoid by entering a period of dormancy. Small birds, with high surface area to volume ratios, are at particularly high risk of lethal heat loss in prolonged cold spells. Annual fluctuations in numbers of Great, Blue and Coal Tits *Periparus ater* correlate strongly with late winter or early spring temperatures, and a good many other species show similar patterns. Survival of Wrens *Troglodytes troglodytes* and Treecreepers *Certhia familiaris* is sharply reduced in cold winters, and the upward trend in the British Wren population since 1965 was sufficiently strong for it to be recognised as a climate change indicator by the UK government. In the last 35 years of the 20th century, the BTO's Common Birds Census (CBC) estimated that the British Wren population increased by over 50 per cent, largely due to improved overwinter survival (Pearce-Higgins 2017). Subsequent BTO surveys, the Breeding Bird Survey (BBS) and BirdTrack, have furthered our understanding of bird responses to climate change. The sensitivity list extends beyond passerines: Lapwings *Vanellus vanellus*, Redshanks *Tringa totanus*, Grey Herons *Ardea cinerea*, Barn Owls *Tyto alba* and Kestrels *Falco tinnunculus* are among those in which survival rates increase demonstrably as a function of warming winter temperatures.

Barn Owls are susceptible to bad weather and have suffered in cold springs despite benefiting in the longer term from warmer winters.

Black Grouse are adapted to cool moorland habitats but their chicks are vulnerable to food shortage in cold springs.

Reproductive success is another important population parameter, and this often depends on weather in spring and early summer. Cold or unusually wet conditions at this time of year increase chick mortality, but in terms of overall population effects this seems to be less important than winter temperatures for 'altricial' passerines – those that raise immobile young within a nest. For 'precocial' species with mobile chicks, including many that live on or around water, the story is different. Growth rates of, among others, Golden Plover *Pluvialis apricaria*, Redshank and Lapwing chicks all correlate with temperature, as do the reproductive outputs of Common Sandpipers *Actitis hypoleucos*, Dunlin *Calidris alpina*, Mallard *Anas platyrhynchos*, Wigeon *Anas penelope* and Black Grouse. Chicks of these species forage for themselves and are therefore vulnerable to exposure and food limitation in much the same way as are adult birds in winter. Raptors are 'semi-altricial', with partially mobile chicks that can suffer from cold springs, as has become evident for Hen Harriers *Circus cyaneus*. In Scotland, cold and wet weather lengthens the brooding time and reduces food acquisition by male Hen Harriers, resulting in low chick survival rates.

Do population sizes of long-distance summer migrants arriving to breed in the UK also benefit from warmer weather? For some, at least, climate in their overwintering grounds and along their travel routes is more important than conditions in their summer quarters.

In particular, survival and abundance of birds such as Sand Martins in and around the Sahel in North Africa correlates with the wildly fluctuating rainfall in that region. Wet conditions in the semi-desert zone generate greater food supplies than are available in dry years, and thus support higher survival rates of migrants passing through them. Once here, though, good weather benefits many migrants just as it does the resident birds. Unfortunately, the positive influence of warmer temperatures on life-history traits does not necessarily translate into population increases. Black Grouse in the UK are in continuous decline due to habitat degradation and hardly constitute a success story. A similar situation pertains for many migrants, including Swallows and House Martins. Just as with all the rest of British wildlife, human activities mediating habitat change or loss are in most cases more important determinants of population size than anything the climate can offer.

Sand Martins (pictured above in Berwickshire) are also affected by habitat and climate on their migration routes; the Sahel region of semi-desert in North Africa for example (pictured below) has highly variable rainfall.

It would be wrong to imagine that climate change is offering universal benefits to British birds. For one thing, warming is not always beneficial. Summer drought is problematic for species that depend on soft, wet ground to supply food such as earthworms and craneflies. The insects abound in good years and can be taken as either subterranean larvae or as hatched adults. Members of the thrush family prey heavily on these invertebrates, and several of these bird species have declined substantially in the UK in recent years.

The Ring Ouzel *Turdus torquatus* is a case in point. Gilbert White was fascinated by the transient visits of this distinctive, white-collared blackbird at Selborne each spring and autumn. Knowing them to be an upland species, he was nevertheless intrigued by their transient appearances: 'What puzzles me most, is the very short stay they make with us; for in about three weeks they are all gone.' Like most people of his time, White was not a sentimentalist. In the case of Ring Ouzels, culinary delights were on a par with the joys of an observant naturalist, and he goes on to remark that 'I dressed one of these birds, and found it juicy and well-flavoured.' Ring Ouzels are summer visitors that breed in the highlands of Wales, south-western and northern England, Ireland and Scotland. They have declined sharply since 1979, mainly as a result of reduced adult and juvenile survival. Breeding sites have shifted uphill, and there may be a damaging effect

The Ring Ouzel is a summer visitor that may be suffering from drier and warmer summers in its upland habitat.

of warming weather on their food supply at lower altitudes. This is by no means proven, and other factors including habitat quality and events at their overwintering sites in the Mediterranean region may be of greater significance. Nevertheless, Devon naturalist John Barkham strongly suspects that climate has played a part in the decline of these birds in a county where he has enjoyed their annual arrival for decades past. As he comments:

> *On my home patch of Dartmoor, the numbers of breeding Ring Ouzels, Curlews and Dunlins have steadily declined. One Ring Ouzel always used to sing close to one of my research sites on the moor; another nested in a former quarry; now neither no longer. Dartmoor is one of their most southern breeding grounds in Europe I think. And the Dartmoor habitat is almost unchanged in 40 years.*

Ongoing research will hopefully get to the bottom of why this charismatic visitor is on its uppers.

As mentioned earlier, there is at least one other reason why climate change could be bad news for some birds. Changes in the relative phenologies of predator and prey species, such as the hatching of chicks and caterpillars, could lead to mismatches sufficient to impact on the predator's population dynamics. Although the evidence is equivocal, something of this kind might be contributing to Pied Flycatcher declines. In northern Europe, the MLDs of this migrant have advanced by as much as eight days, in response to warming temperatures. However, in many places the peak abundance of their main caterpillar prey has advanced by much more than that, and the flycatchers now breed too late to coincide with the chicks' food supply. Fortunately for the birds, caterpillar population dynamics vary according to woodland type. Their peak abundance is much sharper in forests dominated by deciduous trees than in those with mixed deciduous and coniferous species. In mixed forest sites caterpillars are available for longer, and flycatchers have increasingly moved into mixed forest habitats. Pied Flycatchers have declined in much of Europe, including the UK, but since 2002 they have recovered unexpectedly in the Netherlands, probably as a result of this shift in habitat preference. Also in that country, long-distance migrants have suffered more marked declines than resident species or short-distance migrants, in accord with the observation that birds arriving from far away are the least able to advance their MLDs.

Breeding in mixed woodland, such as here in the New Forest, is allowing Pied Flycatchers to accommodate changes in the phenology of insects upon which their chicks depend.

Overview

Just as with plants and invertebrates, some vertebrates appear to have adjusted the timing of their breeding behaviour to compensate for warming temperatures. Phenological responses have been best documented in amphibians and birds, largely because these are the animals most easily watched at the critical times. For freshwater fishes, reptiles and mammals, data deficiency limits our understanding of their phenology, though there are strong hints that many of them, too, have advanced their reproductive activities. It would be particularly interesting to learn more about the mammalian reproduction, for example whether Red Fox *Vulpes vulpes* or Badger cubbing times have advanced as winters have warmed. Bats, too, deserve more attention. Do they still hibernate for as long as they did in the past, or have start or end times of dormancy, or both, changed to reduce the hibernation period? It is easy to imagine that milder winters could work against bats, and perhaps other mammals, by reducing metabolic slow-down at a time when little food is available. Late winter and early spring are critical periods for many animals because food production is still limited, and emerging from torpor with low reserves of body fat is a recipe for disaster. This does seem to be an issue for some amphibians, and it might be a more widespread downside of mild winters than is yet realised.

Again, just like the situation with plants and invertebrates, there have been winners and losers as a result of climate warming in the UK. Among freshwater fishes and amphibians there has been very limited evidence of climate-related changes either way in terms of distribution or abundance. Some warm-water river fishes have extended their ranges upstream and one non-native amphibian has improved its prospects for long-term persistence, but not much else has been detected. Sand Lizards, on the other hand, are showing preliminary signs of niche extension and may be benefiting from warmer summers whereas, just possibly, the opposite may be the case for Viviparous Lizards. With mammals there is little news, but the UK seems to be acquiring an extra bat, and Badger proliferation might be at least partly climate-driven.

There is sound evidence for alterations in biogeographical range and improved population dynamics as consequences of climate change in a good many species of birds, though it is often hard in particular cases to distinguish between climatic and other factors as primary causes of change. Damaging effects of phenological mismatches between nesting times and prey abundance have happened, but the extent of this problem is far from clear. Factors other than chick survival are often more critical determinants of population size, and it has sometimes proved possible for birds to change or extend habitat selection to obtain the necessary food supplies. As with many flying insects, water is an ineffective barrier to movement, and birds can readily extend their ranges across open sea. A positive impact of climate change has been the recent establishment of several new birds as regular breeders in the UK, a trend surely destined to continue.

Temperature changes may affect length and timing of hibernation in bats.

Fungi, lichens and microbes

chapter five

The fruiting bodies that delight us by emerging unheralded on autumn lawns, in waysides and woodlands, and as imposing outgrowths on tree trunks, are what most of us recognise as attention-grabbing fungi. Lichens, those curious adornments of tree branches, drystone walls, house exteriors, gravestones and indeed almost any solid surface apart from plastics, may appear fundamentally different but in fact are also underpinned by fungi, and are a curiously attractive element of British wildlife. Other microbes, mostly bacteria, maintain a (usually) invisible but well-known presence in all habitats on land and in water. It is important to understand what impact, if any, climate change is having on these ecologically vital organisms.

A word about the title of this chapter, which is somewhat tautological. Lichens are mostly made up of fungi, and some microbes, such as yeasts, are also fungi. Nevertheless, for naturalists there is a clear distinction between all of these important and often beautiful groups of organisms. The UK is home to several thousand species of large fungi and around 2,000 varieties of lichens, but these numbers are changing upwards all the time and not much attention need be paid to these estimates. New species are being discovered for real, or created by splitting existing ones on the basis of DNA evidence, at an astonishing rate. Microbes, including bacteria, some fungi and taxonomically contentious groups of protists (unicellular organisms such as amoeba), number in the tens of thousands at the very least, and again their diversity continues to grow as molecular techniques reveal ever more varieties.

One reason to dedicate a chapter to these organisms is the huge importance that fungi and bacteria have in decomposition and nutrient

OPPOSITE PAGE:
One of autumn's bewitching spectacles: fungi on a woodland floor in the New Forest.

recycling, without which no ecosystem could persist. However, it is the contribution of fungi to the pleasure of an autumn walk that really matters. Lichens are there to delight us all year round, but the story is rather different on the rare occasions when we think about microbes. Many are individually very beautiful under the microscope, but we are more likely to encounter them as problems: perhaps as a bloom of foul-smelling blue-green algae in a polluted pond or lake, or as pathogens wreaking havoc among our treasured woodland trees. We also have to reckon with viruses, which are not microbes but usually get an even worse press. Some pathogenic varieties may be spreading as a result of climate change, though the overwhelming majority of viruses infect microbes and are just another component of a healthy ecosystem. It is just as well for pond dippers that they are mostly harmless, as natural freshwaters typically contain upwards of ten billion virus particles per litre. So while we should avoid unfair judgements on what are for the most part valuable components of all ecological communities, there is increasing concern that a minority of nasty microbes may be aided by climate change. We need to know more about that, as well as what might be happening to the good guys, including of course the large fungi and lichens throughout the land.

Fungi

Phenology

The association of fungal extravagance with autumn is a longstanding and justifiable one, and this is the main season for fungal forays by those with a primarily culinary interest in these eye-catching adornments of the forest floor. Of course fungi do not only appear in the autumn, and some species are conspicuous at other times of year. Most obviously, many of the bracket fungi found on tree trunks are always present, and there are also ground dwellers including morels like *Morchella esculenta* which in the past routinely fruited only in spring. However, a significant consequence of climate change for fungal phenology has been a widespread collapse of regular fruiting times. In Peter Marren's words:

> *On fungi, one mourns the loss of predictability. One of the reasons I gave up leading fungus forays is that you can't rely on a decent showing of mushies any more, not even in October.*

The changing habits of fungi have no doubt arisen due to increasingly indistinct transitions between the seasons. Towards the end of the year, in particular, the gradient of falling temperatures from late summer through to winter is much shallower than it used to be. One consequence of this change is that the window of opportunity for fungal fruiting under ideal conditions of temperature and moisture has widened dramatically. Marren (2012) makes this point with the benefit of many years' experience, and once again we are indebted to dedicated amateur enthusiasts, including members of the British Mycological Society, for information on this score.

It is surely a wonderful testament to the UK's long tradition of natural history that every group of organisms has its band of committed followers. Outstanding examples in this case are Ted Gange and Tim Hindley, who accumulated a database of fungal fruiting times sufficient for a powerful statistical analysis. Starting in 1950 and continuing over no fewer than 55 years, these men generated 55,000 records of 2,000 species of fungi at 1,400 sites in south Wiltshire and the New Forest. Among their revelations were that some species now fruit earlier than was usual in the past. One example of this trend, Fly Agaric *Amanita muscaria*, can now appear any time from June to November. The average length of the season for many fruiting bodies has extended from a previous norm of about a month, peaking in early October, to 75 days, which in mild autumns can run up close to Christmas. Spring fruiters have also responded to mild winters, sometimes to an

Morels are spring-fruiting fungi that often appear a month or so earlier than was previously normal.

Fly Agarics are among the mushrooms that have a greatly extended fruiting season, and they can now be seen from midsummer almost until Christmas.

LEFT: False Morels are spring emergents that have advanced their fruiting time remarkably, showing above ground as early as February.

RIGHT: Fairy rings such as those produced by champignon fungi can delight us in gardens as well as in the wider countryside.

impressive degree. The False Morel *Gyomitra esculenta* now typically emerges from the subsoil in February, whereas 20 years ago April was the normal time. Unsurprisingly, given comparable studies with many plants and animals, the degree of earliness observed in spring fruiters correlates with temperatures in late winter and early spring months.

Another recent development, additional to extended and sometimes earlier fruiting times, has been extra fruiting, in spring, of some wood-decaying species that previously appeared only in the autumn. This has been interpreted to mean that longer growing seasons of trees, coupled with milder winters, have generated increased rates of leaf and wood decay.

The idea is that this, in turn, provides fungi with more nutrients, promoting greater vigour and thus more extensive growth. Fruiting bodies are very much the minor component of a soil-dwelling fungus, the hyphae of which extend in vast networks underground, sometimes for tens or hundreds of square metres. In our Sussex garden a 'fairy ring' of the Fairy Ring Champignon *Marasmius oreades* slowly expanded to a diameter of around 10m in as many years, radiating out from a decaying tree stump. Fungi can live a very long

Fungi, lichens and microbes

Bacteria go unnoticed and unconsidered by most naturalists for most of the time, for the obvious reason that they are rarely visible to the naked eye. Individual bacteria are of course never visible, but exceptions occur when they aggregate to make conspicuous, usually unappealing conglomerate masses. This mostly happens in freshwater, and anyone with an interest in pond life will very likely have come across some of them. Loosely attached filaments of rust-coloured iron bacteria, commonly *Leptothrix ochracea*, are frequent inhabitants of heath and moorland pools poor in organic content. In these nutrient-deficient waters they survive by virtue of a curious metabolism whereby they obtain energy by converting ferrous iron to the orange-coloured ferric form. Sometimes iron bacteria thrive so well that they create continuous orange carpets along the bottom of moorland streams.

Members of the *Nostoc* genus of photosynthetic cyanobacteria, organisms otherwise known as blue-green algae, occasionally form grape-sized, gelatinous globules that stick to water plants and, at least in my garden ponds, appear unpredictably in winter or spring. Apparently some species are a delicacy in the Far East, but they are yet to appear on the menu in British restaurants. Appetising they are not. Other species of cyanobacteria are responsible for mats of scum that sometimes form on ponds and lakes in springtime, especially in polluted ones, generating an unprepossessing appearance and a correspondingly awful smell.

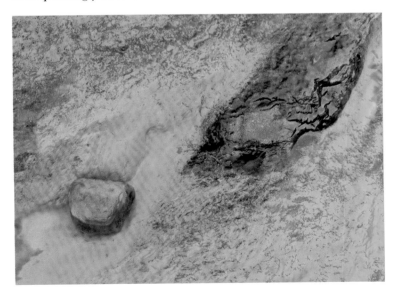

Bacteria are super-numerous in all habitats, but are rarely as obvious as these iron bacteria in nutrient-poor streams.

Are any of these microbes responding to climate change? The rare occasions in which bacteria come to our notice as aggregates serve as reminders of the enormous abundance and diversity of those that we don't see, particularly in soils, where they underpin ecosystems the planet over. Significant effects of climate change on these hidden communities could have profound consequences for the organisms that we *can* see.

Attempts to find out whether anything sinister is going on have, so far, mainly employed experimental approaches for which microbes are often well suited. Growth and species compositions of cultures from soil or water samples can be monitored under conditions that compare the present with future climate change scenarios. Naturally, the value of any experimental results will depend on how good our predictions of those scenarios turn out to be, about which there is more discussion in Chapter 9.

There are other questionable aspects of this approach. Most soil microbes are unculturable as pure strains, and competition and differential survival on a petri dish lead to microbial communities in the lab. This is a serious downside that more than compensates for the ease with which microbes can be manipulated. Differential growth patterns will very often affect the outcome, even in experiments designed to mimic natural substrate conditions as far as possible.

But to end on a high note, a field experiment in California has given cause for optimism about the resilience of bacterial communities in soil. Rainfall was manipulated to simulate climatic variation in parts of a grassland habitat over five years, and its effects on bacterial diversity were measured using molecular techniques capable of identifying vast numbers of species present in soil samples without ever seeing them under a microscope. This is a powerful, DNA-based method, which gets over the experimental problem of unculturability as well as the virtually impossible task of distinguishing one sausage-shaped microbe from another under the microscope. The upshot of the study was that the bacterial community remained the same over those five years, but there was a hint of community changes after that. Of course we have no idea whether any of this matters from the perspective of the fungi, plants and animals that share that grassland with the microbes that live beneath it.

Pathogens

The British countryside is an endearingly benign place for ramblers and naturalists. Any risk of picking up an awful infectious disease while out and about is minimal compared with that in many other countries, including some not far from home. Rabies has been kept at bay, and malaria, once endemic as the 'ague' in medieval fenland, faded out long ago for reasons still unknown. The mosquito capable of transmitting the malarial protozoan, though not the most severe form of it, is still alive and well in the UK. There are fears that if warming continues this disease could eventually return to the British Isles, but there is no sign of that happening yet. Southern Europe remains free of this debilitating infection at the time of writing, so any climate-based threat to Britons cannot be imminent.

There are a few other nasty microbes out in the wilds of the UK, but it is not difficult to avoid them. Pond enthusiasts should be aware of the potentially lethal Weil's disease, caused by a bacterium excreted in Brown Rat *Rattus norvegicus* urine, but you would have to drink pondwater or expose an open cut to it for any significant chance of infection.

Lyme disease, caused by another bacterium, is more problematic, and several thousand people contract it every year in the UK. It is transmitted by biting Sheep Ticks *Ixodes ricinus* that live mainly on large wild mammals, mostly deer. Unfortunately, the ticks regularly detach from their primary host and loiter on grass or bushes from where they readily transfer their allegiance to passing humans. In America the spectre has been raised of climate change exacerbating an already extensive Lyme disease problem, because mild winters improve tick

Sheep Ticks are bloodsuckers of vertebrates and can transmit Lyme disease.

survival, increase egg production and ultimately raise tick population density. This might be happening in the UK too, as infection rates are certainly on the rise, but the jury is out with respect to any proven connection with climate warming. Maybe more significant is the enormous and ongoing increase in British deer populations, the primary wild hosts of this tiny parasite. Perhaps these animals are also suffering – and the ticks don't move across just to humans. On heathlands near forests with plenty of deer it is not uncommon to find several ticks attached to Viviparous and Sand Lizards, with unknown consequences for individual mortality or population viability. In this instance, eating more venison might be a more practical solution for reducing disease risk than controlling climate change.

For humans, then, our countryside remains for the most part a pretty safe place. Some of our wild plants and animals are not so lucky, and new pathogens are arriving on British shores at a seemingly increasing rate. Could climate change be playing a part in all this? Dutch elm disease, caused by a fungus but spread by a beetle, was one of the first of the recently arrived contagions; it appeared in the 1960s and has since destroyed more than 60 million elm trees in the UK. There is a suggestion that warmer and wetter summers speeded up the spread of the disease on the Isle of Man, but no strong link with climate has been implicated elsewhere. Ash dieback, caused by another fungus, *Hymenoscyphus fraxineus*, is one of the most recent visitors. First seen

Blackened, dead leaves still hanging on the tree: tell-tale signs of Ash dieback in a woodland in Norfolk.

in Suffolk in 2012, it threatens to wipe out another very widespread and cherished member of our woodlands, and within a few years the fungus has managed to spread across much of England. Again, though, there is no direct evidence to suggest climate change was a key mover. Probably the arrival of new insects that attack trees, rather than the pathogens, has been a more directly attributable consequence of climate change. The Horse-chestnut Leaf Miner moth mentioned on p. 92 is a prime example, but it is certainly possible that both insect and pathogen damage to trees might be exacerbated by extra stress due to extremes of rainfall and drought.

Animals too have suffered from alien diseases. Among the most devastating is squirrel pox virus, now recognised as a major factor in the catastrophic Red Squirrel *Sciurus vulgaris* declines across much of the UK over the last century. Introduced by, but harmless to, North American Grey Squirrels *Sciurus carolinensis*, this disease has been a critical player in the displacement of the native by the alien species. North America was also the source of crayfish plague, the fungus that has slaughtered native White-clawed Crayfish over most of their range in England following its accidental introduction, again as an accessory, this time with non-native Signal Crayfish in the 1970s.

There are other unpleasant microbes doing the rounds with slightly less devastating but still serious effects. One causes trichomonosis, a debilitating protozoan infection of Greenfinches and some other birds, which has caused widespread population declines. These diseases are those that, so far, have had the greatest impact on native British animals – but in no case has their spread been attributed to climate effects.

This declaration of innocence does not, however, extend to another fungal disease that has been credited as the greatest cause of species extinctions and declines across the globe. The chytrid fungus *Batrachochytrium dendrobatidis* has rampaged through amphibian populations all round the world, and is currently present in most British species. Fortunately, the chytrid has not had detectable pathogenic effects on any British amphibian and seems to be carried as a passive infection. This happy situation is far from true elsewhere. In parts of Europe, Common Toads have been decimated and hundreds of victims have

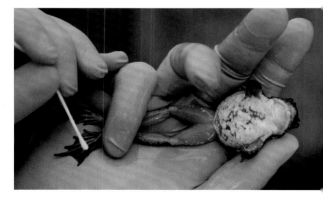

A frog being swabbed to test for the amphibian chytrid fungus in King's Canyon National Park, California.

littered breeding ponds in the Pyrenees. Species vary in susceptibility, with many acting as unaffected carriers, but environment also plays a role in disease outcome, with major mortality events most frequent, in Europe, at high altitudes. Chytrid-mediated mortality of overwintering Midwife Toad *Alytes obstetricans* tadpoles in high mountains is exacerbated by warm winters in which lake ice melts early, though the mechanism connecting early melt with increased virulence is not known, as discussed by Clare *et al.* (2016). Climate change is causing these lakes to thaw progressively earlier year on year, implying that chytrid-induced death rates will also increase. Midwife Toads are not native to the UK but, on the basis of mortalities in the Pyrenees, British Common Toads living at high altitudes could face a similar risk. The odd thing is that Common and Midwife Toads living at lower altitudes in Europe, where water temperatures are warmer than in the mountains, are inherently resistant to the same chytrid fungus that many of them carry around. Until this inconsistent effect of temperature at different altitudes is explained it is premature to anticipate any general impact of climate change on the epidemiology of this unpleasant pathogen.

The effects of another fungal infection, much less serious than chytridiomycosis, may also be influenced by a warming world. *Saprolegnia* species are common British water moulds that grow as white networks of hyphae on dead animals in ponds but that also, in some conditions, can infect and kill live ones. Victims are wide-ranging, from aquatic insects to fish, and include amphibian eggs. *Saprolegnia* is at its most virulent in cold water, where it seems to gain a growth advantage over ectothermic animals and can cause high levels of mortality. A cold spell in spring often signals high *Saprolegnia* infection rates of amphibian eggs, including those of Common Frogs and Natterjack Toads. Perhaps climate change is working against this pathogen now, but direct evidence remains to be collected.

Domestic as well as wild animals may face increased risks from pathogens as the climate warms. Survival of the agent responsible for bovine tuberculosis (bTB), *Mycobacterium bovis*, outside a host animal varies considerably according to environmental conditions. Although its persistence on farmland varies according to soil type, vegetation structure and various other local factors, bTB survival in the field is much enhanced over winter if temperatures remain above freezing point. As most transfer of *M. bovis* into cattle from wildlife vectors probably occurs in pastures, this observation is of some importance because it implies that the trend towards milder winters might exacerbate the impact of an already devastating disease. From the wildlife perspective, if current attitudes persist, more disease might lead to even more persecution of Badgers.

Bovine tuberculosis is not the only problematic pathogen farmers have to cope with. Another, in this case recent arrival is the bluetongue virus that infects both wild and domestic ruminants but is particularly serious for sheep. Until recently this was a disease of the tropics, but since the turn of the millennium it has spread north through Europe, reaching the UK in 2007. The virus is transmitted by biting midges, and it is their increased survival, due to warming winters, that is credited with the northward spread of bluetongue. The virus has found new midge vectors in Europe. In Africa, it is spread by *Culicoides imicola*, which cannot survive European winters, but its new allegiances are with *C. obsoletus* and *C. pulicaris*. Even these insects fare poorly in frosty conditions, hence the link between the spread of bluetongue and climate change that has reduced the intensity of midge winter-kill. This virus might yet impact on wildlife as well as domestic livestock, since deer are susceptible to it and some 40 per cent of Red Deer in Belgium now show signs of exposure to bluetongue.

Beautifully symmetrical but deadly for sheep: a bluetongue virus depicted under the electron microscope.

Overview

Such impact as climate change has made on the fungi and microbes considered in this chapter has mainly involved alterations in phenology, sometimes advancing spring and summer events and in other cases just making them less predictable. There are fewer examples of changes in distribution or abundance, though this may reflect a lack of information for many species rather than the true national situation. Some pathogens may be benefiting from a more comfortable environment, but the UK's most virulent wildlife diseases are managing to wreak devastation on their own without any help from climate change. When climate does work to extend the range or impact of pathogens it normally does so by advantaging the insect vectors that carry them, rather than by acting directly on the organism itself. Arguably it is habitat damage and deterioration, including atmospheric and freshwater pollution, that have dominated the recent fates of fungi, lichens and microbes in the British countryside.

Freshwater and terrestrial communities

changes related to increasingly mild winters have also been detected. In Windermere – a well-studied habitat mentioned earlier in the context of Perch phenology – phytoplankton (algae) blooms, zooplankton (mostly *Daphnia*) blooms and the hatching of Perch eggs have all advanced to peak earlier over the past 40 years, but this did not happen synchronously. Zooplankton bloom-time changed the most, and fish egg hatch-time advanced the least. It is not yet clear what effects, if any, this asynchrony will have on community structure in future, since the species composition across the food web has remained the same.

A different but equally disconcerting observation is that warmer winters seem to be combining with eutrophication to increase the dominance of duckweeds, floating plants that can cover ponds and ditches completely and deprive submerged species of light. This leads to declines or losses of oxygenating macrophytes, with consequent impoverishment of species diversity in the underwater community. Nutrient enrichment looks like the main cause of this effect, but eutrophication is not always the culprit when macrophytes die off. In a Swedish lake with no evidence of changed nutrient load, submerged vegetation declined after 30 years of increasingly mild winters. The most likely explanation was increased overwinter survival of fishes, which caused excessive sediment disturbance and nutrient release, in turn creating unfavourable conditions for the establishment of rooted, submerged plants in spring.

Swedish winters are more severe than British ones, but something similar could be happening in high-altitude lakes in Wales, the Lake District or Scotland. In Windermere, Arctic Charr have declined commensurate with increases in Pike and Roach *Rutilus rutilus*, two fishes that prosper in relatively warm water. Roach are a recent introduction to the lake, highlighting how unregulated fish movements can add extra stresses to species sensitive to other environmental pressures, in this case climate change. Even another cold-water fish, the Brown Trout, becomes competitively superior to Arctic Charr as temperatures rise. Cumbria's fabulous lakes, spreadeagled beneath their magnificent mountainous backdrops, are all change beneath the surface. A combination of eutrophication and warming water is, within just a few decades, transmuting fish communities that must have survived unscathed since the end of the last glaciation.

Some amphibian communities may also be responding to climate change. The phenological disconnect between Common Frog spawning times, which have become a little earlier in recent decades,

Freshwater and terrestrial communities

Pike thrive in moderately warm conditions and are increasing in northern lakes. They may threaten cold-water species such as Arctic Charr.

and the much earlier arrivals of newts after migration to the same breeding ponds, could be serious for frogs. In the past frog spawn usually hatched before most newts appeared on the scene, allowing tadpoles to disperse ahead of the predators' appearance. Now there is much greater overlap and newts can home in on immobile, helpless frog embryos still embedded in spawn jelly. In my garden ponds I have watched serried ranks of newts surround frog-spawn clumps and devour every embryo within a couple of days. Over the space of a few years the frog population declined fast, and others with garden ponds have witnessed similar assaults, sometimes resulting in complete loss of a frog population. This change of fortune may

Different rates of phenological change are increasing newt predation of frog spawn in some ponds.

187

be exaggerated in small garden ponds compared with larger ones in the wider countryside that usually have more varied topography, lower newt densities and shallow spawn refuge areas. Nevertheless, the garden microcosm shows how community dynamics can change quickly when circumstances conspire.

Rivers and streams have not been widely studied in the UK with respect to possible climate effects on their flora and fauna, but we do know that temperatures are on the rise in running water just as they are in lakes. In some Welsh streams, the biomass of large invertebrates including mayfly and stonefly larvae declined coincident with water warming, but in this case there was another complication. The watercourses, now of good quality, had previously suffered from acidification, which reduced biodiversity and maybe increased vulnerability of the surviving species to climate effects. Increasing stream temperatures in Iceland were followed by greater moss cover and a changed invertebrate community, with fewer chironomid midges and more simuliid flies and snails. As recounted in Chapter 4, in some well-studied French rivers the more warm-adapted fishes, such as members of the carp family, are extending their distributions upstream into previously colder waters dominated by salmonids. It would be interesting to know more about the extent to which community changes of this kind are happening in the UK.

In summary, evidence of climate change affecting British freshwater community compositions is real but sketchy, and in most places seems minor compared with the devastating consequences of eutrophication. However, there are hints of more to come if warming continues, and fish communities in particular seem at risk of substantial changes, to the disadvantage of cold-water species.

Mountains and moorlands

The uplands of northern and western Britain are the closest landscapes we have to wilderness in the UK, though only small areas around mountain-tops come anywhere near to being pristine habitats little impacted by human activity. These are also places where concerns about negative impacts of climate change are acute. Species of plants and animals adapted to cool conditions at high altitude are potentially at increased risk of destructive competition from those moving uphill from warmer climes. By nature of their demanding physical features, survey of what is going on in high mountains is

of lichens. Observations were made across nine habitat types: *Calluna* heath, *Vaccinium* heath, *Racomitrium* heath, fell-field, snow-bed, springs, grassland, *Juncus* grassland and *Nardus* grassland. Averaged across all locations, species richness of all three floral groups (higher plants, mosses and lichens) increased in most of these habitats. Once again, as on Ben Lawers, there was also a small but significant increase in average height of vegetation across all habitats and locations. Higher plant species richness increased most markedly in snowfield sites, while bryophytes increased most in fell-field habitats. Only in the highly exposed fell-field zone was there no increase in higher plant richness, a result attributed to extreme wind exposure in these places.

There were, however, significant differences in patterns of enrichment among the four regions. Changes among higher plants were greatest on Mull and in the Cairngorms, whereas lichen richness increased in the Cairngorms but decreased in the other areas. Bryophytes increased everywhere to similar extents. The composition of each habitat group had become more homogeneous over the years, and the largest changes in species composition were seen in springs and at snow-bed sampling points. As with the Ben Lawers region, it was possible to identify winning and losing species on the basis of changed frequencies between the survey periods. Across all four regions the

Acid grassland appears to be thriving but is subject to invasion by new species.

patterns were broadly similar and concurred with what has happened on Ben Lawers, though in the wider study Snow-bed Willow was decreasing. Nineteen species, mostly arctic-alpines, declined whereas ten increased, in this case mainly temperate species with widespread distributions. Changes occurred in both the frequency of individual plants and in their extent of ground cover, not always in the same direction. For the main winners and losers, however, frequency and ground cover shifted in the same way. These strong responders are the plants most likely to underpin future community-level shifts in composition. This amounts to a lot of possible community changes in the coming years.

Relatively low, so-called 'closed' habitats with extensive plant cover were as sensitive to change as the higher 'open' habitats with more bare ground. This was an unexpected conclusion, as greater resilience against community change was expected in the closed habitats. Springs, snow-beds and *Vaccinium* heaths showed the largest increases in plant diversity, while fell-field proved the most stable habitat. Similar changes, often even more marked, have been reported in upland areas of Wales and England, but in these places eutrophication is a more significant complicating factor. Scotland is not free from nutrient enrichment, but, especially in the high Cairngorm mountains, there is less evidence of it than in highlands further south. Eutrophication may, however, be partly responsible for lichen declines in the other three Scottish regions compared in the Britton investigation.

Snowfields in the Cairngorms are home to rare plant communities suffering significant changes as the climate warms.

Freshwater and terrestrial communities

The evidence of adverse impacts from a warming climate on high-altitude plant communities is there to see, and concerns about it seem fully justified, as described by Britton *et al.* (2009). Apart from being an ecological disaster in the making, we could lose some strikingly attractive plants. On the at-risk list are Trailing Azalea, with its bell-shaped pink flowers, white-bloomed Starry Saxifrage *Saxifraga stellaris*, Cowberry *Vaccinium vitis-idaea* with its shiny, pea-sized fruits, Snow-bed Willow and a whole banquet of fetching lichens and mosses including Witch's Hair Lichen *Alectoria nigricans*, Madame's Pixie-cup Lichen *Cladonia coccifera*, Racomitrium Moss *Racomitrium heterostichum*, Silky Forklet-moss *Dicranella heteromalla* and Dusky Fork-moss. Stiff Sedge is retreating, while Carnation Sedge is advancing. Ling Heather is the only colourful flowering plant set to increase in abundance; most of the other winners are mosses such as Rusty Swan-neck Moss *Campylopus flexuosus*, Fir Clubmoss and Pohlia Moss *Pohlia nutans*, as well as Whiteworm Lichen. It looks as if bryophyte and lichen communities are the ones undergoing the most marked responses to climate warming.

LEFT: The arctic-alpine species Starry Saxifrage, one of the upland plant community's losers.

RIGHT: Ling Heather, here in the Lammermuirs, is benefiting from climate change in the uplands.

BELOW: Whiteworm Lichen is an arctic-alpine species that seems to be prospering despite climate change.

About possible changes in montane animal communities there seems to be little known, although some cold-adapted invertebrates have declined or are moving further uphill as life gets warmer, as discussed in Chapter 3. These include several species of butterflies, moths and dragonflies as well as at least one beetle. Vertebrate diversity is low in the high mountain regions, but there are charismatic species including Mountain Hare *Lepus timidus*, Ptarmigan *Lagopus muta* and Snow Bunting *Plectrophenax nivalis*. So far, all these animals remain incumbent and their communities are unchallenged, though concerns are regularly raised about their long-term prospects.

Mountain Hares may suffer if snowfall decreases sufficiently to reduce the camouflage value of white coats in winter.

Snow Buntings, like Mountain Hares, are not yet responding detectably to climate change but could become vulnerable in future.

Ancient woodland, as here in the New Forest, is relatively resilient to climate change, but there have been alterations in the flora of woodland understoreys.

Woodlands

Although the UK has less woodland than most of its European neighbours, forest cover is increasing, and across the islands there is a great diversity of forest types. Pine-dominated remnants of the Caledonian forest in Scotland are very different communities from the mixed, mainly deciduous, woodlands of lowland England. There have certainly been many changes in British woodlands over the last half-century, but the challenge is to discover which of them, if any, have been influenced by climate change. Woodlands have rich faunal as well as floral communities, and as prime habitat for many birds they have repaid investigations of possible dislocations between advances in nesting times and abundance of insect prey.

Anyone going down to the woods today, if it was their first visit in a long while, would be sure of a big surprise. The total extent of broadleaved woodland, most of which is in England, has increased by 50 per cent since the 1940s but this reflects a lot of new planting and masks a substantial loss of ancient woodland. Scrub and coppice have virtually disappeared in most places, and now over 95 per cent of woodland is dominated by high, canopy-covering forest. Inevitably this has created big changes in the understorey, with less light penetrating to the forest floor and resultant declines of plants and invertebrates associated with open ground. Pollution has been less important than in some other habitats in driving change, because broadleaved trees are less bothered by increases in nutrient levels, and acid rain proved to be a transient problem now much reduced. Most of what has gone on in recent decades is due to more direct human actions, including reductions in coppicing and ever-increasing use of spinneys and copses as covers for game. The unregulated releases of enormous numbers of Pheasants *Phasianus colchicus* would not be permitted for any other non-native species. Predatory impacts of Pheasants and the herbivorous excesses of escalating deer numbers, now out of control in all too many woodlands, have had very noticeable effects on ground flora and fauna. Reptile populations have been wiped out by Pheasants along forest rides, and Bluebells have been trashed by Reeves Muntjac Deer *Muntiacus reevesi* in East Anglia, threatening one of the most joyous spectacles of a British spring.

Despite the dominating interventions listed above, British woodlands have not been untouched by climate change. Trees have, as yet, not been seriously affected, but there is growing concern about the future prospects of Yew and, especially, of Beech woodland. Beech trees are very susceptible to drought, and the dry summer of 1976 caused mass die-offs in the New Forest and elsewhere. At Lady Park Wood in Gloucestershire, Beech trees were still dying 41 years after the drought event. Growth rates are reduced by periods of water stress, and if climate change predictions about drier summers prove correct we may see declines in the contribution of Beech trees at least to forests in lowland England. Sue Everett expressed her concerns about this future threat based on what she saw in Italy:

A Beech tree in Lady Park Wood that was badly scarred by the drought in 1976.

High-altitude Beech forest looks like leaves emerged early then leaves killed by cold. Obviously there will be big impacts from wide-scale killing of young Beech growth. Very hot weather expected soon, so soil will be exposed as canopy dies. I suspect there will be heavy epicormic growth [buds sprouting from beneath the bark] of leaves but trees with small tight canopies that are close together will no doubt die back significantly. Collared Flycatcher food source will be greatly depleted, plus other knock-on impacts.

However, it is among the understorey ground flora in woodlands that the strongest signals of climate effects are visible so far in 103 woodlands spread across Britain, according to Kirby *et al.* (2005). The longer growing season consequent on earlier springs and later autumns has affected the extent of ground cover by 17 out of 65 species examined. Thirteen of these suffered reductions, including Bell Heather *Erica cinerea*, Common and Early Dog-violets *Viola riviniana* and *V. reichenbachiana*, Wood-sorrel *Oxalis acetosella*, Honeysuckle, Wood Avens *Geum urbanum*, Bracken *Pteridium aqulinum*, False Brome *Brachypodium sylvaticum* and Tufted Hair-grass *Deschampsia cespitosa*. The decline in Bracken was surprising, since this fern generally benefits from mild, frost-free winters. No pattern was discerned to suggest any differences among the responding species with respect to early or late growing or flowering, or in relation to British distribution patterns. Furthermore, 51 out of 332 species sufficiently well documented for statistical comparisons showed correlations between frequency of occurrence and winter/early spring (January through to March) mean temperatures between 1961 and 1999. Not all species were suffering. Species expanding most markedly included Heath-grass *Danthonia decumbens*, Common and Pill Sedges *Carex nigra* and *C. pilulifera*, Harebell *Campanula rotundifolia*, Selfheal *Prunella vulgaris* and Wood Anemone *Anemone nemorosa*. Again it was hard to detect any ecological pattern that characterised these beneficiaries of a warming climate, though there was a trend for them to prefer wetter springs and drier summers than non-responding plants.

Dog-violets (left) are losing out as climate change impacts the flora of the woodland floor, whereas Harebells (right) are doing increasingly well.

There is of course more to woods than vegetation. Woodlands have been among the communities most thoroughly investigated by animal ecologists, including attempts to discover whether phenological advances, induced by climate change, have generated mismatches in the timing of critical events in breeding cycles. Disconnects could arise at various trophic levels, maybe between plant and herbivore, or between predator and prey. In either case, starvation or food limitation might impact on the population dynamics of any species involved. In spring woodlands, bud-burst often signals the start not only of plant growth but also of the attack of herbivorous insects that feed on them. Usually the herbivores in question are the larvae of insects rather than the adults, and eggs must be laid so that caterpillars hatch at exactly the right time for food to be available. The effects of climate change on both plant and herbivore can be complex because, for both contestants, warm winters might simply advance both activities or they might act discordantly if, for example, the plant or insect requires a period of cold (vernalisation) before bud-set or oviposition. In such situations the trend might be for a delay in bud-burst but an advance in the appearance of the insect, or vice versa, either of which could have serious consequences for the herbivore.

There are possible ways around this problem. The insect eggs might have an ability to delay hatching time, even if laid early, until tender new growth begins and generates a hatch stimulus. Alternatively, the insect could simply track bud-burst and wait for the right time to lay. For example, the jumping plant louse *Cacopsylla moscovita* is, like many insects in that family, highly host-specific. The nymphs feed on catkins of Creeping Willow *Salix repens*, and the female lays her eggs between developing catkin ovaries. Because the larvae need the entire catkin development period to mature, it is vital that eggs are laid as soon as catkins are available and that the eggs hatch almost immediately. The female louse can store her eggs for at least six weeks before laying them, and this ability serves as a buffer against delayed catkin appearance. This kind of adaptation seems rare, though, and in this case has probably been selected for because the event timing is especially crucial.

A particularly well-studied system involves bud-burst in oak trees and egg-hatch in Winter Moths *Operophtera brumata*, the larvae of which feed on oak leaves. The timings of both events are influenced by winter and early spring temperatures, and in both cases are advanced by increased warmth. Bud-burst and egg-hatch have advanced to more or less the same extent, so in this case climate effects are unlikely to change the interactions between eater and eaten. Experimental studies testing whether differential effects of vernalisation on the plant and insect might confuse the apparently simple synchronisation failed to find any differences, confirming the stability of this trophic interaction. However, at higher trophic levels there are other important aspects to this ecosystem. Blue Tits and Great Tits rely heavily on moth caterpillars to feed their young, and once again timing is critical. Egg-laying by both birds has advanced in many localities, just as it has for the moths, and for the birds coincidence of these breeding patterns is essential if the chicks are to fare well. Chicks that hatch too early or too late, relative to the peak of caterpillar abundance, are likely to starve. This possible dilemma was recognised early on by climate change watchers, and has verged on becoming a paradigm for the ecological consequences of asynchrony.

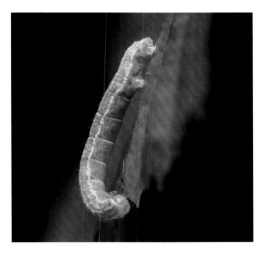

Winter Moth caterpillars constitute a major food source for the chicks of many woodland birds.

The tit and caterpillar question has been investigated in the UK, but some of the most detailed work has been carried out elsewhere in Europe. Tits have advanced egg-laying times concordant with climate change at Wytham Woods in Oxfordshire (Hinks *et al.* 2015), but at Hoge Veluwe in the Netherlands, Great Tits failed to change their dates of egg-laying between 1973 and 1995 even though temperatures were generally on the rise, and peak caterpillar abundance advanced by ten days over this period. Why the regional difference? It all seems

Great Tits are one of many avian predators of woodland insects that might be affected by changes in caterpillar phenology.

to depend on exactly when temperatures were changing. Over the egg-laying period (March to mid-April) there were significant temperature increases in the UK but not in the Netherlands between the 1970s and 1990s. However, there was a trend towards increasing warmth over the subsequent month when the chicks needed food. Presumably this discrepancy between early and late spring temperature trends confounded a more synchronous response, since post-laying climate can hardly be anticipated by the birds.

There were significant adverse consequences for the tits when caterpillar peak abundance was out of step with chick development. Second broods became rarer, fewer chicks fledged, and their average weight was lower than when synchrony was maintained. However, adult survival across years was not significantly affected by food and brood asynchrony in spring, and there were only weak effects on population demography overall. Annual variation in the mean number of new recruits per female, which is the main determinant of population fluctuations in Great Tits, was driven mostly by population density and by variations in the autumn Beech crop. It seems that the extent of mismatch between caterpillar peak and tit chick development varies in space according to local differences in the details of how climates are changing, and at some sites synchrony between chick and caterpillar hatching has been well maintained (see figure below). More importantly, so far there has been no strong negative effect of this well-studied phenological asynchrony at the bird population level, as shown by Reed *et al.* (2013). Further up the food chain, Sparrowhawks have failed to advance their breeding time to correlate with tit fledging,

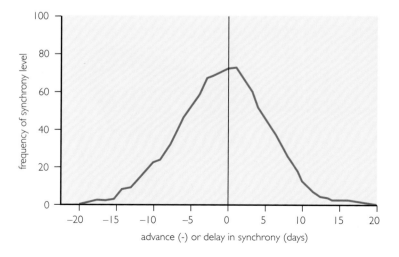

Synchrony of Great Tit chicks at ten days old and peak of caterpillar production between 1999 and 2010 in Finland. Asynchrony varied in both directions, with chicks hatching up to 20 days before or after caterpillar peak hatches, but with synchrony (0 day difference) the most common situation. Adapted from Pakanen et al. (2016).

generating another level of mismatch. Once again, though, no adverse consequences have been noticed at the population level for this top predator. On the contrary, Sparrowhawks have been fairing increasingly well in much of Europe in the 21st century.

Tits and caterpillars are not the only species combination that has the potential to impact on bird communities. There has recently been an interesting disconnect between the breeding activity of Cuckoos and their unlucky hosts. As for most long-distance migrants, Cuckoo arrival times have become only slightly earlier than in the past, advancing by about one day per decade since the late 20th century. However, the resident species that have commonly been parasitised by Cuckoos, including Dunnocks *Prunella modularis* and Meadow Pipits *Anthus pratensis*, have been laying earlier than before, generating a mismatch in their favour. A reduction in the proportion of nests of these species invaded by Cuckoos has followed. To make up for this deficit there has been a concomitant increase by the parasite in the use of nests of a long-distance migrant, the Reed Warbler *Acrocephalus scirpaceus*. These changes in host selection have been greatest in areas with the most rapid spring warming, and could change the relative abundances of susceptible species and therefore the composition of bird communities over time.

Reed Warbler feeding a Cuckoo chick. The warblers have suffered increased parasitisation by Cuckoos, as other potential victims are now breeding before Cuckoos arrive.

Dunes and heaths

Coastal dunes and lowland heaths share a few common attributes. Both form on well-drained soils, mainly sands and gravels; and both, because rain percolates quickly through the substrate, tend to be nutrient-poor, although this is more marked for heaths than for dunes. Sand dunes receive inputs of wind-blown shell calcium from the neighbouring sea but are inherently unstable owing to winter storms and associated gales. Both habitats support mainly low-growing vegetation and therefore become very warm at ground level in summer. Flora and fauna are distinctive and different from those

Sand dunes of the coastal fringe are on the front line against increasing winter storms.

Lowland heaths, a warm habitat becoming warmer still.

of other habitats in the UK, though heathland vegetation, dominated by heathers, bears resemblance to that of upland moors. Dunes and heaths are home to a wide range of specialist plants and invertebrates and are the best places to find reptiles. Until recently, many coastal sand-dune systems were among the most pristine habitats in the UK, little modified by human hands. By contrast, heathlands were created by humans, starting several thousand years ago, and their wildlife has been modified by changing management practices ever since, but especially over the past century.

Although dunes and heaths support ponds, bogs and, occasionally, streams, they are for the most part dry places that in warm summers become decidedly arid. This makes them potentially vulnerable to climate change. Many of the pools on these habitats are shallow and temporary, drying up even in an average summer. Warmer and drier conditions would make them even more ephemeral or remove them altogether, endangering specialists such as Tadpole Shrimps *Triops cancriformis* and Natterjack Toads that rely on them for successful breeding. Terrestrial vegetation can become tinder-dry, and, especially on heathland, it is very prone to devastating wildfires. Drought reduces the competitive superiority of Ling Heather over invasive grasses such as Wavy Hair-grass *Deschampsia flexuosa*, which may compound the impact of nitrogen deposition, an increasingly important influence on the vegetation of both heaths and dunes. The combination of eutrophication and climate change makes post-fire recovery of heathland vegetation increasingly likely to diverge

Temporary ponds are at risk from premature desiccation if summers become warmer and drier.

Bracken is a frost-sensitive plant that spreads rampantly on heaths and moors but seems to be retreating in some woodlands.

from the historically normal succession back to heather, in favour of grass invasion. The further spread of Bracken, another scourge of lowland heaths, is also encouraged by warmer weather, especially mild winters.

Animal communities on heathland are somewhat volatile, because a few species on the edge of their biogeographical range in the UK only survive here at all on account of the warm microclimate. If warming continues, these species may feature in heathland communities more strongly than they have in the past. The best-known contender in this respect is surely the Dartford Warbler *Sylvia undata*, which almost disappeared from the UK after the severe winter of 1962/63 but has since spread, episodically, from Dorset to heaths as far north as East Anglia. However, this success might be short-lived because habitat specialists are declining relative to increasingly abundant generalists across the country as a whole (see the Birds across the UK section, p. 211).

Extreme weather events have become more frequent in recent years, almost certainly because the warmer atmosphere generated by climate change contains more moisture than it did in previous decades. The consequential increase in storminess poses a danger to coastal dunes, threatening to create massive erosion, loss of sand and destruction of vegetation. It is not clear, though, whether this type of trauma will become more significant than it was historically, nor whether it will inevitably be a bad outcome. British dune systems have undergone periods of instability before, interspersed with calmer conditions, in what may be longstanding put poorly understood cycles. On Merseyside there was massive dune erosion at the end of the 19th century. Aerial photographs taken shortly afterwards show what looks like a desert landscape almost devoid of vegetation. By the mid-20th century these same dunes were very different, with little bare sand and overgrown with dense stands of vegetation. It is hard to know to what extent the current vista is a part of a natural cycle and how much has been influenced by nitrogen deposition and the spread of a nitrogen-fixing plant, Sea-buckthorn *Hippophae rhamnoides*, which is a recent introduction to that area.

Merseyside dunes, highly eroded, in the 1930s (top); and the same dunes, over-fixed, today (bottom).

Repeat surveys of vegetation at 89 coastal sites in Scotland, in 1976 and 2010, revealed no differences that could be attributed to climate change, although during the intervening 34 years rainfall increased by about 5 per cent, average summer temperatures by about 0.8°C, and winter temperatures by 0.6°C. However, on Ireland's west coast dune fixation increased dramatically and rapidly at 11 widely separated locations between 1995 and 2005. Bare sand exposure decreased by between 51 and 96 per cent on these dune systems, and correlated with increased numbers of growing days due to earlier springs and later autumns, all as described by Jackson & Cooper (2011). Ireland's Atlantic-facing coastline is less vulnerable to nitrogen pollution than are seaboards in Britain, and in the Irish situation climate warming could be the main contributory factor underlying the floral community changes. Whatever the truth of the matter, conservationists are increasingly interested in destabilising many of Britain's coastal dune systems to improve the future prospects of a specialised flora and fauna diminished by over-fixation. Perhaps more storminess might be advantageous to this ambition.

Although dunes and heaths certainly appear vulnerable to climate change, there has as yet been little in the way of evidence indicating serious climate effects in either habitat. Perhaps this is because the most serious threats to these environments, the warmer and drier summers predicted by climate change models, have been slow to materialise. Heathland fires occur and can devastate wildlife over large areas, as they did on Thursley Common, Surrey, in 2006. About 60 per cent of the heath was lost in that conflagration, including almost all its Sand Lizards. This is not a new problem, though it is probably increasing in frequency. Gilbert White described devastating fire damage at Woolmer Forest, in his day a completely open heath, in the 18th century. Then and now, most heath fires have been started deliberately by humans – livestock farmers in the past, arsonist vandals today. Vegetational changes, too, are mainly the result of human actions, be it nitrogen pollution loads or altered management. Climate change, by comparison, has been a minor player in the recent fate of these hugely important wildlife habitats.

Wildfires on heathland, such as this one at Ashdown Forest in Sussex, are an ever-increasing problem exacerbated by warm and dry conditions.

Agricultural land

Peering out of an aircraft window while approaching Gatwick airport, all looks well with Blake's green and pleasant land. A panoply of fields, hedges, copses, ponds and lakes still extends indefinitely into the distance as a glorious patchwork that would have been familiar to generations of our forebears down on the ground. Perhaps that is why so few politicians appreciate the serious plight of our countryside and its wildlife in the 21st century.

The aerial impression is an illusion. Much of the UK's farmland, by far the most extensive of the country's habitats, has become a wildlife desert since the Second World War. The primary cause, industrialisation of agriculture with its unfettered applications of fertilisers and pesticides, is all too obvious to anyone looking more closely than from an aeroplane cabin window. Given the intensity of these chemical and physical assaults, is climate change of any relevance to such a widely degraded habitat?

Farmland is of course not uniform and traditionally included arable fields, hay meadows and pasture. Until the mid-20th century, these land uses were of low intensity. Hay meadows were usually grazed in winter, while arable crops were commonly rotated from year to year, approaches that incidentally favoured wildlife as well as optimising income for the landowner. Lowland hay meadows have almost disappeared from the UK, and their restoration is the subject of great efforts by conservation organisations. The diverse meadow plant communities, once riots of summer colour across the land, are rare and precious now and, fortunately, the survivors are considered resilient to climate change. Upland hay meadows are more likely to alter, as plants such as Smooth Lady's-mantle *Alchemilla glabra* and the glorious Wood Cranesbill *Geranium sylvaticum* are already being wrested from their preferred climate envelope.

Smooth Lady's-mantle, an attractive inhabitant of northern hay meadows that is vulnerable to climate change.

Detailed investigations of climate impacts on farmland communities are few, but a long-term study of arable land in Sussex, running from 1970 to 2011, attempted to disentangle climate and pesticide effects on the wildlife of cereal fields and was reported by Ewald *et al.* (2014). Almost half of the invertebrate groups investigated were sensitive to hot, dry summers or cold, wet winters, and trends in invertebrate numbers over the 40-plus years correlated with those of temperature or rainfall. Abundances of spiders, leafhoppers, adult bugs, thrips, braconid wasps and mould beetles increased in hot and dry years and decreased in cold and wet ones. Planthoppers, rove beetle larvae, silken fungus beetles and fungus gnats increased in both hot and dry, and cold and wet years. However, increasing use of pesticides was more important as a driver of wildlife changes than any climate effects.

Unsurprisingly, most of the interest about climate change on farmland has dwelt on farm productivity and the possibilities of growing different crops, rather than on concerns about wildlife. There is little to suggest that the animals and plants surviving under modern agricultural regimes have much to fear from climate change, at least in the short term, which is a sad indictment of the state of most of the UK's countryside today. The risks are low because so little is left. Long-term prospects may be more complicated. Many plant-eating insects, including moths, butterflies, beetles and bugs, often do not utilise the full geographical ranges of their food plants, implying that other factors limit their distributions, as described by Stewart *et al.* (2015). Climate warming can permit the spread of insects into areas previously suitable for the host plant but not for the consumer, but also into places with different plant species that the insect is now able to consume. This niche broadening can change community structure because new competitive interactions, potentially destructive, can arise with insect species already present.

Aside from the parlous state of its wildlife, open country is another arena for some postulated disconnects in phenology. As in woodlands, these concerns have mostly been unrealised so far. The average emergence times of Orange Tip butterflies have tracked the first flowering dates of one of its host plants, Garlic Mustard *Alliaria petiolata*, remarkably closely, as both events have become earlier in response to warming March temperatures.

Arguably of greater significance would be serous mismatches between pollinators and the pollinated. A clever approach to this question compared events reaching as far back as the 19th century with what is going on today. The early information was derived from herbarium records of the Early Spider-orchid *Ophrys sphegodes* and museum specimens of the Buffish Mining Bee *Andrena nigroaenea*, all as described by Robbirt *et al.* (2014). For both species the historical records included dates on which the specimens were collected. Almost all pollination of this orchid is carried out by male bees of this single species, deceived into pseudocopulation by the flower's mimicry of the female insect. Both organisms have advanced their phenology, but the bee by a greater degree than the plant. Field observations revealed that between 1975 and 2009 *Andrena* flight time correlated with early spring temperatures, and in warm springs they advanced by as much as seven days. Pretty much the same relationship with temperature held for bees collected between 1893 and 2007 when their catch times were compared with historical climate records. Male bees fly earlier than females, on average by about four days. However, whereas orchid flowering times have advanced by more than six days per 1°C of March–May temperature rise, male bees advanced their flying time by around eight days. Female bees flew earlier too, but by less than males, and consequently appeared more or less coincident with the orchid flowers. Orchids compete with female bees to attract males, and the changed phenology means that the orchids no longer have the males 'to themselves' before female bees appear. This is potentially serious for plants in which seed production is pollinator-limited. Currently, however, the status of Early Spider-orchids seems to be stable, following declines in the early 20th century. Perhaps the libido of male bees is up to coping with the new situation.

This Early Spider-orchid (left) will almost certainly have been pollinated by a male Buffish Mining Bee (right), mistaking it for a mate.

Evidently phenological mismatch could have significant consequences for highly specific pollinator-plant couples where only one or a few insect species do the job. Most situations, however, are not like that, and when there are many pollinators available there is correspondingly less chance of serious impacts of mismatch on the plant population. When phenology has advanced it is almost always the case that the plants and insects bound by pollination have responded to the same cues of increased spring temperatures, thus minimising the chances of substantive dislocation. Presumably for this reason, most studies of this potentially important subject around the world have shown maintenance of synchrony between pollinator emergence and flower availability, and though there are a few exceptions no plant decline has yet been attributed to pollination failures. As the orchid example shows, however, there is no room for complacency – and it is likely to be some of our rarest and most cherished species, with highly specific pollination requirements, that will be the first to suffer if asynchrony continues to widen.

Rewilding

The idea of recreating wilderness of a kind not seen in the UK for thousands of years has attracted increasing interest from conservationists. Several groups are promoting the idea of creating rewilded areas, each of at least 100,000 hectares, in which wildlife would be free to live entirely unmolested by human activity. An interesting aspect of the concept, not without controversy, is the proposed reintroduction to these places of predators long extinct in the British Isles, including Lynx *Lynx lynx* and Wolves *Canis lupus*. There are few places in the UK where this amount of territory could be made available, so the main foci for rewilding would very likely be in the Welsh hills and Scottish uplands. Some less dramatic developments, on a much smaller scale, are already under way, including the Great Fen in East Anglia. In no cases, though, are these existing operations left completely free of human intervention. If it ever happens, rewilding will create wildlife communities quite unlike any of the ones considered above.

Apart from benefiting wildlife, another claimed justification for rewilding is that it would increase resilience to climate change by creating areas to sequester carbon and to absorb floodwater. 'Rewilding Scotland and tackling climate change for Christmas' is the

The Great Fen is a long-term project aiming to restore wetland to benefit wildlife and people.

way forward, urged Trees for Life (www.treesforlife.org.uk), going on to make the point that 'every tree dedicated will help reduce the impact of climate change by replacing the carbon footprint and packaging of Christmas presents with a gift that instead soaks up carbon dioxide, and that benefits wildlife and Scotland's wild landscapes.' Perhaps rewilding is an idea whose time has come. I hope so.

Birds across the UK

One result of climate change is that, like most other animals, many birds in Europe are moving northwards. This has consequences not just for the individuals involved but also for the communities in which they settle. Between 1994 and 2008, changes in the distribution of around 260 species were tracked in the UK. The results were remarkable. There was a general increase in species diversity in all the habitats where counts were made, notably on farmland including arable, improved grassland, mixed farmland and natural grassland, in woodland, in urban sites and in upland regions, as shown by Davey et al. (2012). This was the expected result consequent upon some new arrivals, moving in from the south, mixing with longstanding residents. Relative to 1994, diversity increased most markedly in the uplands, where it was inherently low, followed by natural grasslands, and was minimal on arable land.

ABOVE: Carrion Crows are among the 'generalist' birds thriving in the face of climate change.

RIGHT: Dartford Warblers have benefited greatly from warmer winters but are among the 'specialist' birds that may be at long-term risk from increased competition by successful generalists.

This sounds like good news, but there is more to it than counting the number of species. What has actually happened is that 'generalists' that thrive in a wide range of habitats have done increasingly well, whereas 'specialists' with less catholic tastes are being eased out. The UK is therefore supporting more common species while rarities are being put at risk. Generalists and specialists are titles of convenience and do not form discrete groups, but are scored by the BTO on a continuum-of-points basis. Among the 'most generalist' on the BTO list are Buzzard, Carrion Crow *Corvus corone*, Chaffinch *Fringilla coelebs*, Kestrel and Siskin *Spinus spinus*; among the 'most specialist' are Cetti's Warbler, Dartford Warbler, Pochard *Aythya ferina*, Snipe *Gallinago gallinago* and Little Ringed Plover *Charadrius dubius*. While species diversity initially increases across the board, this trend goes into reverse as specialists retreat and bird communities become increasingly homogenised in species composition.

By 2008, the biggest increases in absolute species diversity compared with 1994 were in East Anglia. However, when the changes in specialist relative to generalist components of bird communities were quantified as a 'community specialisation index' there were substantial declines in the specialist contribution across much of south-east England, the Midlands, most of northern England, Wales, and in Scotland south of the Great Glen. Temperature changes were negatively correlated with the specialisation index, implying that the process was at least partly climate-driven.

Nevertheless, caution is once again required, because other factors are certainly at work to underpin these changes. Farming practices, for example, are surely the primary reason for the fact that only small increases in relative diversity were reported on arable land. The underlying trend in much of the UK is for declines in abundance of most farmland and many woodland bird species.

Overview

Climate-driven changes of community compositions, including any consequences of phenological mismatches, have, thus far, been minor in most parts of the UK. The high mountains are an exception, and these include places where floral species mixes are altering on a worrying scale. Other habitats show less evidence of changes consequent on warming conditions, but those that might be occurring are likely to be masked by complicating factors, especially eutrophication and agricultural intensification.

This pattern of little apparent effect in most habitats is to be expected, bearing in mind that substantive climate change has only kicked in over the past 20–30 years, a brief period compared with the lifespans of many plants and animals. Only species that can adapt rapidly to new conditions, such as annual plants and insects, and perhaps the highly mobile birds, are likely to respond quickly enough for early detection of significant change, and that has been the story so far. This implies that communities may be less stable than they currently appear, and that more extensive changes may well be on the cards if climate warming and rainfall patterns continue as predicted in future decades. As usual, though, climate impacts on wildlife communities currently pale by comparison with habitat deterioration.

The open sea around St Kilda is part of a vast underwater world all around the British Isles, with rich and diverse communities beneath the waves.

south from the Arctic. It is the latter, nutrient-rich, source that has in the past supported high biomasses of plankton and the fishes that feed on them around our shores. Water in the southern reaches of the North Sea is mostly no more than 50m deep, sufficiently shallow for complete mixing from top to bottom of the water column when the water is cold. During the winter, strong winds ensure that this actually happens, liberating nutrients that precipitate a phytoplankton bloom as the water warms in spring. Stratification follows, with phytoplankton in the warm upper layers initiating a transient flush of zooplankton. This in turn attracts young fishes, especially juvenile Lesser Sand Eels, close to the surface. Historically this process generated a glut of accessible small fishes at exactly the right time of year to supply seabirds that were busily nesting and laying eggs. Sand eels spent the winter on the seabed and returned to it in early summer as the stratification of upper sea levels intensified, nutrient concentrations diminished and the plankton declined. However, very young sand eels remained near the surface long enough to provide food for seabird chicks. This critical food web has long underpinned the huge seabird colonies of the UK's coasts and offshore islands.

Climate change has created a situation in which the demarcation between the influence of warm and cold water has moved steadily northwards. The North Sea, in particular, has warmed by an average of 1–2°C since 1980. The geographical situation of this waterway has rendered it particularly susceptible to temperature change. Constrained between Britain to the west and the European mainland to the east and south-east, warm water from the English Channel can move north through the Straits of Dover into the southern, shallow section of the North Sea that was dry land until about 8,000 years ago. The 10°C marine isotherm in this region has moved northwards at approximately 22km per year since the 1960s. Some warming also results from Gulf Stream spillover around the north coast of Scotland. The shift is having major impacts on Britain's coastal and marine wildlife. In the Galapagos, El Niño events are occasional and their effects are quickly reversed. Not so in the UK, where there is an ongoing trend for warm water to supersede cooling Arctic influences.

Climate-induced changes in North Sea productivity are now well understood. Nutrient-rich cold water is increasingly displaced northwards and overlaid by a warmer, nutrient-poor stratification. This has interfered with the ecosystem in two ways. Firstly, there is an increasing phenological mismatch between Lesser Sand Eel spawning times, which have advanced, and seabird egg-laying times, which have remained largely unchanged or become slightly later. Warming water has reduced the growth rates of sand eels, partly because they hatch too

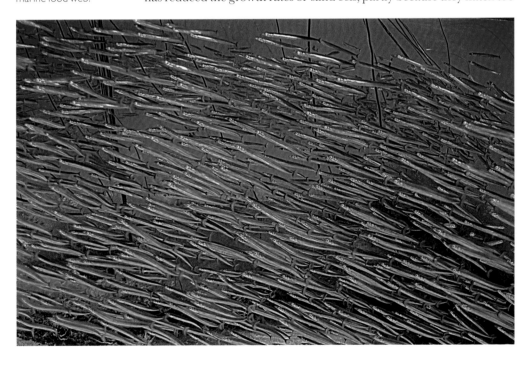

Lesser Sand Eels, vital components of a complex marine food web.

The copepod *Calanus finmarchicus* is a cold-water crustacean declining in the North Sea as the water warms.

soon to match up with the zooplankton peak. Secondly, and probably of greater significance, there have been changes in the zooplankton species themselves and their population dynamics consequent on the northerly spread of warmer water. The copepod *Calanus finmarchicus* is a cold-water member of the zooplankton community that blooms in spring, and in the past it has been a primary food source for small fishes, including sand eels. However, it has been increasingly replaced by the warm-water species *Calanus helgolandicus*. A problem has arisen because the invading copepod population blooms in the autumn, too late to support young sand eels, and this has precipitated declines in springtime copepod and thence fish biomasses.

Sand eels are also having a hard time for another reason. As with some amphibians on land, warmer winters increase their metabolic rates and cause them to use up fat reserves just to stay alive. Lower fat reserves contribute to reductions in young eel growth rates in spring, compounded by reduced plankton availability at that time of year. One crucial consequence of these changes is that sand eels are now both smaller and fewer than they used to be, at a time when demands on them from predatory birds are at their highest.

Further changes in the community composition of the UK's offshore waters are inevitable and, indeed, are already under way. Anemone Prawns *Periclimenes sagittifer*, which live among the tentacles of sea anemones, were first recorded in the UK in Dorset in 2007 but have since been discovered as far east as Selsey in Sussex. This crustacean was formerly found in shallow seas around mainland Europe, but not near the UK.

Another remarkable addition to the offshore flora indicates just how profound future changes might be. The diatom *Neodenticula seminae* has, for the past few hundred thousand years, been native to the subarctic Pacific Ocean. Prior to that it occurred more widely, including in the North Atlantic, but the species died out there more than 800,000 years ago. We know this because, on account of their silica-rich outer cell walls, diatoms persist as fossils for many millennia. Diatoms also have distinctive shapes which mean that species can often be identified in marine sediment core samples. In the 1990s *Neodenticula seminae* reappeared in Nordic seas and rapidly became widespread. This was possible because of the decline of summer sea ice in the high Arctic, opening up the historic 'north-east passage' and thus connecting the Pacific and Atlantic Oceans with marine currents across northern Russia that had not operated since the Pleistocene.

Changing fish faunas

Warming seas have precipitated substantial changes to the marine fishes around British coasts, especially in the North Sea where the temperature changes have been the most marked. Effects on North Atlantic fisheries have been widely investigated by, among others, the International Council for the Exploration of the Sea. There is an interesting historical perspective to concerns about temperature changes in the North Atlantic. During the 1920s and 1930s, a warming trend similar to, though smaller than, that observed recently led to reduced ice cover in the Arctic and to higher sea temperatures. Atlantic Cod *Gadus morhua* and Halibut *Hippoglossus hippoglossus* expanded their ranges northwards while cold-water species such as the Beluga Whale *Delphinapterus leucas* remained further north in winter than they had done in previous colder years. These observations precipitated a scientific meeting about climate change at Copenhagen in 1948. *Déjà vu*, we've been here before, albeit at a time when climate warming was transient and not sustained in the way it has recently become.

Now fish are on the move again. Between 1977 and 2001, half of 20 marine fishes that were monitored extended their ranges northwards, some by tens of kilometres. The Bib (or Pouting) *Trisopterus luscus* was an extreme case, advancing by more than 300km. Sardines *Sardina pilchardus* invaded the North Sea from the

Guillemot breeding along North Sea coasts have suffered from reductions in the Lesser Sand Eel population.

food stock. These include Common Terns *Sterna hirundo*, Arctic Terns *S. paradisaea*, Puffins *Fratercula arctica* and Kittiwakes *Rissa tridactyla* as well as Guillemots, Razorbills and Shags, all as described by Pearce-Higgins & Green (2014).

Some birds have coped with changes in the availability of this prey animal better than others. Despite their predilection for sand eels, neither species of tern has declined dramatically in recent decades, and indeed some populations of Arctic Terns have increased despite decreases around Shetland related to reduced sand eel numbers in the late 1980s. The resilience of terns may relate to their migratory habits, as both species are summer visitors to the British Isles and spend much of the year elsewhere.

Razorbills have declined recently near Iceland, apparently in response to a lack of sand eels, but around the British coasts their populations seem fairly stable. Guillemots are able to dive deeper than many other seabirds and can catch sand eels all year round. Even so, their breeding success in some locations such as the Isle of May in south-east Scotland has depended on the availability of sand eels hatched the year previous to the birds' breeding season. The abundance of this eel age class has crashed since the 1980s, but in many areas the birds have successfully coped with fewer sand eels by dieting on alternative fish species. Overall, in contrast to the situation in some North Sea colonies, the British Guillemot population is faring well and has tended to increase since 2000.

Other seabirds around British shores have not been so lucky, and recent declines have been documented by the Joint Nature Conservation Committee Marine Biodiversity Monitoring programme (jncc.defra.gov.uk). Shags have decreased by over 30 per cent since the 1980s and their chick productivity has varied over that period according to fluctuations in sand eel abundance. Their plight, however, might have as much to do with another consequence of climate change. Mass mortalities known as 'wrecks' following serious storms have been documented on several occasions and appear to be the major factor affecting adult Shag numbers. The likelihood of ever more extreme weather events, including storms at sea, does not bode well for the future of British Shag populations.

Shags are in decline due to a reduced food supply and because of mass storm-generated mortalities.

Puffin declines are well documented, and also relate to the reduced sand eel population on which their chicks largely depend.

Puffins have also suffered from storm 'wrecks', including flooding of their nest burrows, but declines since 2000 also relate to reductions in sand eel numbers. The average masses of food brought by adult Puffins to their chicks on Fair Isle declined by 80 per cent between 1996 and 2015. Attempts by Puffins to broaden the spectrum of their prey have been made but have proved largely unsuccessful. Between 2004 and 2008, some seabirds including Puffins incorporated Snake Pipefish *Entelurus aequoreus* into their diets. Pipefish were uncommon in British waters before 2004 and became so again after 2008, a variation in abundance of unknown cause although their initial increase was originally ascribed to climate change. Unfortunately for the birds, this bony fish has a low energy content and can even choke chicks. The pipefish did not provide a sustainable alternative food source at a time when sand eels and Sprats *Sprattus sprattus* were scarce, and chick productivity declined accordingly. Puffins have declined seriously in some of their main strongholds, especially around Iceland – where they still appear on restaurant menus – and also to an uncertain degree in the UK since the start of the 21st century.

Puffins are in trouble, but Kittwakes are in more of it. This species nests on ledges along vertical cliff faces, mainly in northern Britain and on the surrounding islands, in colonies that can number up to the tens of thousands. Its status around our coasts has diminished

Climate Change and British Wildlife

Kittiwakes have suffered more than any other seabird from the dearth of sand eels.

to a staggering extent since the 1980s, with an overall population decrease of some 60 per cent. Kittiwakes are especially susceptible to food shortages because they can only catch fishes at or near the sea surface, unlike diving species that are able to exploit more of the water column. This restrictive behaviour has exacerbated the effects of low sand eel abundance, as has commercial fishery for this prey item. By 2008, just one chick was fledged, on average, from every four nests, compared with about one per nest in the 1980s. Such low productivity implies that further declines in Kittiwake abundance are likely in future as fewer chicks eventually enter the breeding population. Productivity recovered somewhat between 2009 and 2014, possibly due to higher numbers of sand eels appearing in those years, but in 2015 the situation reverted to one of very low breeding success.

Kittiwakes have the dubious distinction of being the bird on which damaging impacts of climate change are most clearly demonstrable. All of this unhappy trend makes the unique population breeding inland on the Tyne Bridge and nearby buildings in Newcastle, as opposed to natural cliff ledges, particularly important and worthy of conservation. Hopefully pressure to move them on, accused as they are of making too much mess, will be successfully resisted.

Marine mammals and reptiles

British waters are rich in cetaceans and seals. Like seabirds, they form an upper end to the marine food chain and are potentially susceptible to any disruption of it caused by climate change. Nevertheless, the strongest signal of climate effects so far has been a positive one – the appearance around the British Isles of species normally associated with more southern climes, as described by Evans & Bjørge (2013). Short-beaked Common Dolphins *Delphinus delphis* and Striped Dolphins *Stenella coeruleoalba* have moved up around the north coast of Scotland and into the North Sea, possibly in response to the increased numbers of Herrings *Clupea harengus* and Anchovies now found there. Herrings declined disastrously in the 1970s due to massive overfishing but subsequently bounced back and are once again common on fishmongers' slabs. Some recent whale and dolphin sightings, including beach strandings, have been of warm-water species not previously recorded around the British coast. These include Blainville's Beaked Whale *Mesoplodon densirostris*, Fraser's Dolphin *Lagenodelphis hosei* and Dwarf Sperm Whale *Kogia sima*. There have also been multiple strandings of another warm-water cetacean, Cuvier's Beaked Whale *Ziphius cavirostris*, around the west coasts of Britain and Ireland since 2005.

Short-beaked Common Dolphins have responded to warming west coast waters by moving as far as the north coast of Scotland.

Narwhals favour cool water and are seen around the UK with decreasing frequency.

On the other hand, most northerly, cold-water cetaceans have not increased in British waters; rather the opposite, as Bowhead Whales *Balaena mysticetus* have not been seen for more than a century and there has been an increasing dearth of Narwhal *Monodon monoceros* and Beluga reports. Although these trends can be interpreted as climate-related, sightings of cetaceans are mostly casual and lack statistical rigour. It is entirely possible that other factors are in play, perhaps including the changing distributions of prey species, which could of course be climate-related and thus indirect influences on the movements of whales and dolphins. The White-beaked Dolphin *Lagenorhynchus albirostris*, for example, is a northern species that has increased recently in the southern North Sea, a trend that can scarcely be explained by simple climate change models.

Genetic studies have added to the current perspective of climate effects on cetaceans. The Harbour Porpoise *Phocoena phocoena* is a common cold-water species with a wide distribution in the North Atlantic. There is, however, a small and apparently quite separate population off the west coast of Iberia. Genetic analysis strongly suggests that as recently as a few hundred years ago, in the Little Ice Age, the Iberian and North Atlantic populations of Harbour Porpoises were contiguous but were separated as the Bay of Biscay warmed up over the past couple of centuries. This gives an interesting idea of how climate-driven changes in distribution can operate, with quite dramatic results, over slightly longer timescales than those happening now. Harbour Porpoises are seen increasingly rarely along Britain's south coast, implying that a northward distributional shift may be continuing.

Compared with cetaceans, there are more quantitative data on the two seals that breed around British coasts, namely the Grey *Halichoerus grypus*, for which the UK supports perhaps 50 per cent of the global population, and the Common or Harbour Seal *Phoca vitulina*. In recent decades Grey Seals have increased while Common Seals have declined around the UK, but there is no evidence that this relative shift in abundance is climate-linked.

Leatherback Turtles *Dermochelys coriacea*, some of the largest reptiles on earth, are increasingly regular visitors to the UK's offshore waters. This species is remarkably tolerant of cold, North Atlantic conditions because it is partially endothermic and has anatomical adaptations that minimise heat loss. Other species of marine turtles occasionally crop up around the British Isles, but these wanderers are way outside their thermal comfort zone, and are almost inevitably doomed to perish. In all marine turtles, breeding involves burying eggs on sandy beaches in much warmer climes than we have around the British Isles. Leatherback visitors to our shores come mostly from Caribbean nesting sites, and the attractions in mid- to late summer are enormous swarms of jellyfish on which the turtles feed. The appearance of Leatherbacks around British shores is nothing new, and reports of these leviathans

Leatherback Turtles are summer visitors that relish the increasing numbers of jellyfish on which they feed. This one was seen in Wales.

stranded around the British coastline go back to at least the 18th century. Thomas Pennant, one of the UK's earliest naturalists and a regular correspondent with Gilbert White, described the outcome of perhaps the first official record (quoted by Bell, 1849):

The late Bishop of Carlisle informs me that a tortoise was taken off the coast of Scarborough in 1748 or 1749. It was purchased by a family then resident there, and several persons were invited to partake of it. A gentleman, who was one of the guests, told them it was a Mediterranean turtle, and not wholesome; only one of the company ate of it, who suffered severely, being seized with dreadful vomiting and purging.

Numbers of turtle sightings are increasing, especially along the south and west coasts. There seems little doubt that the growing popularity of British shores for turtles is down to increasing numbers of jellyfish swarms, which in turn may be responding to the warming seas. Jellyfish are the main diet of Leatherback Turtles which, because this prey is more than 95 per cent water, have to consume enormous quantities of them. While undoubtedly good news for the turtles, holidaymakers are less enthusiastic – and are regularly urged by newspaper headlines to stay out of British waters because deadly cohorts of jellyfish are heading our way. Fortunately there is also no shortage of advice on how to cope with a sting, should you ignore advice to forgo the pleasure of a swim. Recommended remedies to apply are bread soaked in milk, urine, vinegar or shaving foam. Probably only one of these would be readily to hand on the beach, but apparently razor-blades, shells or credit cards can remove embedded nematocysts.

Jellyfish numbers appear to be on the increase around the British coastline.

Coastal and marine environments

Unfortunately, any positive result of increased jellyfish numbers for turtles may be more than counterbalanced by rising temperatures and sea levels in their nesting areas, both of which have the potential to reduce recruitment to the turtle populations. A threat peculiar to some reptiles, including marine turtles, arises because they have temperature-dependent sex determination, which occurs during embryo develoment in the buried eggs. Rising temperatures have the capacity to alter the sex ratio of offspring, with potentially serious consequences for population dynamics. For a variety of reasons most sea turtles are in global decline, so we can only hope that thriving feeding grounds around the British Isles help one species a little bit.

Thus far marine mammals and the single marine reptile conform to a common trend seen elsewhere in both terrestrial and marine habitats, namely one of increasing species richness, with warmth-loving species sighted ever more often. As yet there is little, but not negligible, evidence that cold-water cetaceans are retreating northwards, and it remains to be seen whether the pattern of increased diversity is sustained over a long period.

Changes in seawater acidity

The world's oceans serve as a sink for carbon dioxide, which is a very soluble gas, and thus act to dampen the impact of greenhouse gas emissions on climate. However, there is a price to pay for this service. Dissolved carbon dioxide forms carbonic acid and thus acidifies seawater to a degree that is increasing concern about its effects on ocean organisms. Acidity is measured as pH, which varies on either side of neutral (pH = 7), upwards to signify alkaline conditions and downwards as a function of acidity. Acidification is measured on a logarithmic scale, such that pH 6 is ten times more acid than pH 7.

Typically, seawater has a pH in the range between 7.5 and 8.5. This mildly alkaline environment supports a chemistry that is favourable for the proliferation of a wide range of animals and plants. It permits, for example, the use of calcium to construct insoluble structures such as snail shells. When carbon dioxide dissolves in pure water, the carbonic acid thus formed can reduce the pH to below 6. This level of acidity solubilises calcium and inhibits the production of solid structures, including shells and corals, which are based on calcium carbonate. Fortunately, most aquatic habitats are buffered by soluble salts that prevent such large drops in pH. Even so, the massive amounts of carbon dioxide produced by human activities over the past century or more have reduced the average pH of seawater by about 0.1 units. Because pH is a logarithmic scale measure, this is a bigger change than it might appear. It translates into about a 30 per cent increase in acidity, and if gas emissions continue to increase unchecked, ocean surface waters will be nearly 150 per cent more acidic by the end of the 21st century than they are now. Such oceanic acidity would be unprecedented in the past 20 million years.

Acidification effects in marine environments have proved difficult to assess in the wild, and most studies thus far have been experimental, imposing pH variation on individual species and on their various life stages in the laboratory. The experiments provide warnings about what may be happening beneath the surface of our seas and oceans, suggesting that acidification can affect a wide range of fish behaviours, via sensory disruptions of smell and hearing, that probably diminish survival prospects. There is little evidence to suggest that small changes of acidity damage adult fish physically, but sperm, eggs and juveniles of some species have proved more sensitive. Despite these gloomy predictions, the experimental method makes many assumptions about future trends in acidification and is inevitably simplistic when looking at individual species rather than at entire food webs.

One enterprising group of researchers took a different approach, exploiting the natural variation in seawater acidity at increasing distances from volcanic activity, all as described by Milazzo *et al.* (2016). Male Ocellated Wrasse *Symphodes ocellatus* create and defend nests on the seabed, and fish activity around these nests was filmed at sites near to volcanic emissions, and at others much further away, all around Vulcano Island near Sicily. Carbon dioxide concentrations were therefore either high near the volcano, as predicted for seas generally in the future, or ambient, close to the current norm, when further away. There were some interesting differences in fish behaviour under the two sets of conditions. In the high-acidification zone, dominant nesting males achieved only about one-third as many 'pair spawns', where just one dominant male accompanied a female, as males achieved in the ambient acidification zone. Dominant males commonly obtained multiple 'pair spawns' by attracting several females, individually, to the nest. Small 'sneaker' males attempted to fertilise eggs at all the nest sites, and sperm competition could be intense. Nevertheless, the lower numbers of pair spawns at the more acidic sites did not lead to higher levels of multiple paternity, as assessed by genetic tests of offspring.

Pictured in 2003, the River Irfon in mid-Wales had become virtually lifeless as a result of acidification from air pollution.

This was a surprising observation, given the ubiquity of 'sneaker' males, and left dominant males as successful in acidified sites as they were in the ambient acidity zone.

This imaginative study provided clues as to how acidification can influence events in the real world, as opposed to in laboratory tanks. However, as a one-off it inevitably gives no more than an interesting hint about bigger pictures that might be emerging as oceans acidify. Beyond European shores, relatively mild acidification is already having some ominous consequences. As part of the global trend, seawater in the Olympic National Park in Washington State, USA, suffered a reduction in pH from above to below 8 between 2000 and 2016. Almost certainly consequent upon this acidification, the average shell thickness of California Mussels *Mytilus californianus* dropped by almost 30 per cent, from 6.7mm to 4.8mm, between the 1970s and the 2000s. This increased shell fragility hardly bodes well for the future of California Mussels, and may well reflect what is happening to other species with high calcium requirements.

To those familiar with the damage done to freshwater streams and lakes by acid rain during the 20th century, the significant impact of apparently very minor pH changes in seawater seems surprising. Even the worst-case scenarios leave seawater in the mildly alkaline range, above pH 7. Base-poor, un-buffered freshwaters of mountain lakes and moors in Europe and North America were acidified tenfold

Upper saltmarsh, a coastal habitat under threat from erosion, together with its unique wildlife communities.

or more, often down to pHs of less than 5, during a period of gross atmospheric pollution. At that level of severity many ecological consequences followed, including declines of invertebrates, amphibians and fishes. Indeed, some sites in soft-water regions of northern Europe lost all their fishes. Like climate change and its primary causative agent carbon dioxide, acid rain was also caused by the release of damaging gases, in this case mostly sulphur dioxide. Happily, acid rain has diminished, and many ecosystems have largely recovered following changes in the emissions of power stations and other polluting industries. Remission and recovery from atmospheric pollution is therefore possible, but unfortunately trying to remove carbon dioxide from fossil-fuel burners is technically more difficult than reducing the release of strongly acidic gases.

Overview

Climate change has influenced the wildlife in the UK's offshore waters every bit as much as it has done on land. Warming seas have changed marine ecosystems and, at least temporarily, increased species diversity within them, just as warmer air has brought new animals, and allowed others to move further north, on the mainland. As with terrestrial flora and fauna, there has also been a downside. Saltmarshes are under threat around the British coastline, eroded and squeezed by rougher seas, higher tides and rigid sea walls.

Arguably the most negative effect of climate change anywhere in and around the UK, matched only by the plight of arctic-alpine plants, has been the disruption of North Sea food webs and the consequences of that dislocation for seabirds. It is hard to find any other British animal that has suffered from climate change as much as the Kittiwake. The enormous seabird colonies on our offshore islands, along coastal cliffs and stacks, and occasionally on city blocks, are among the most dramatic spectacles that British wildlife has to offer. Not just Kittwakes, though they seem to be in the greatest difficulty, but other seabird species too are threatened by warming waters and diminishing stocks of prey. How dreadful it will be if the downward trends continue, and we are eventually left with barren cliff faces resounding only with the crash of pounding waves.

Acidification by increasing concentrations of carbon dioxide is a problem mostly associated with the marine environment, and one with ramifications that are very poorly understood. The peculiar response of Ocellated Wrasse breeding behaviour to elevated acidity could not have been predicted. It may be unique and perhaps of little consequence even to the fish concerned, but it certainly illustrates how little we know about the future impact of acidification. Effects on marine snails and corals are already evident and will be an increasing worry.

No doubt carbon dioxide concentrations are also rising in many freshwater habitats, and this issue is now receiving attention, though the worst damage that carbon dioxide can do to them is far less than what acid rain achieved. In the absence of any buffering, pH from dissolved carbon dioxide cannot fall below about 5.6, and even this level of acidity will only occur in relatively pure water. The sea and many freshwaters have buffering salts that moderate acidification from carbon dioxide such that they cannot fall anywhere near as low as pH 5.6. In some extreme cases during the 1980s, acid rain reduced pHs in poorly buffered freshwaters to less than 4. Apart from during these brief and hopefully unreturning conditions of recent history, over longer timeframes many freshwater organisms must have experienced greater variations in acidity than those currently living in the sea. Hopefully this will make freshwater organisms relatively resilient to future carbon dioxide increases, but only time will tell. What is clear is that for marine animals susceptible to acidification, the only possible relief will come from control of greenhouse gas emissions.

drawn by Sands, from a Picture by Zoffany, R.A.

ROBERT MARSHAM ESQ. F.R.S.
OF STRATTON STRAWLESS, NORFOLK, OB.T 1797.

been influenced by the publicity and, all too often, by damaging personal experiences. The floods in northern England during the winter of 2015/16 were the worst on record, devastating almost 6,000 households. However, massive inundations are not new. Once again 1947 emerges as a dramatic year. In that spring, owing to high rainfall and snow-melt, the Thames valley experienced its greatest flood of the 20th century. Among other observations of that calamitous event, around London it was noted that the floodwater was approximately 60cm deep in the Bexley Arms but that the Vansittart Arms, slightly to the north and on a rise in the road, was not affected. Pubs were clearly a matter of high priority. There was apparently some singing, by marooned residents, of the 'Eton boating song', to keep spirits up.

But have extreme storm and rainfall events actually increased in frequency or intensity? Are they true indicators of climate change? Records show that storminess has increased in the UK over recent decades but not to an extent greater than in the 1920s. This of course was before the experiences of the overwhelming majority of people alive today, making the recent trend obvious and easy to link with climate change. Nevertheless, because of the longer-term data set, there is no hard evidence to confirm such a link. The Meteorological Office has consistently emphasised that individual extreme climate events in the UK cannot safely be charged to global warming. Despite the caveats, though, it is now broadly accepted that heavy rainfall episodes are on an upward trend. As Professor Myles Allen of Oxford University proclaimed, 'armchair meteorologists who continue to insist this is all just weather are starting to sound a little bit like Aunty Mabel expressing surprise at her remarkable luck in board games.' This will be ongoing bad news for those living in high-risk areas, where effective flood defences will become ever more impractical. Rather sad, then, that government recommendations about not building houses on flood plains continue to be widely ignored. As for wildlife, long-term trends in butterfly and bird populations have apparently not been much affected by individual extreme weather events although they have coincided with occasional population crashes and, more rarely, with population explosions in moths and birds, as revealed by Palmer *et al.* (2017).

Working outside

Farming

The overwhelming majority of people in the UK live and work in urban areas. With relatively little exposure to the great outdoors, ongoing effects of climate change probably pass many urbanites by, although the abundance of gardeners (see p. 253) may make this too simplistic an assumption. A very different situation pertains to farmers. Although agricultural workers constitute only about 1 per cent of the British workforce, they are responsible for managing some 70 per cent of the UK's land surface. Farmers naturally have a huge vested interest in weather, as crop yields in summer and livestock survival in winter are at its mercy. If any one group of people is likely to have noticed the impact of Britain's changing climate, it is surely the farming community. However, despite its obvious significance for farming practices and economics, surveys have shown that many people working on the land do not recognise climate change as a major threat, as determined by Wiles (2012). Therefore most farmers, until recently, were unwilling to invest time or resources in tackling its likely consequences. They need to be persuaded, because farming is contributing significantly to climate change. In the UK 8 per cent of greenhouse gas emissions comes from agriculture, mostly as nitrous oxide and methane released from fertilisers, livestock, farm machinery and the loss of organic matter from soils.

Responding to climate change is certainly in the best interests of farmers. Farming has changed almost beyond recognition over the past 70 years, driven mostly by increased mechanisation, fertiliser and pesticide use. In the 1950s oats were the dominant cereal in the UK, whereas now wheat, barley and oilseed rape are the main arable crops in much of the country. This major evolution of the landscape has had nothing to do with climate change. On the other hand, problems with fruit production are clearly linked to climate. Warmer winters have reduced the incidence of chilling, an essential requirement for full development of some overwintering crops and fruits. Orchards, already greatly diminished in number and extent, may be further reduced by this problem, at least in southern England. Summer droughts have increasingly impacted yields of brassicas and tomatoes in parts of the country.

As time goes by

Extremes of weather have adversely affected livestock production: high temperatures reduced milk output in the summer of 1990, and the unusually cold spring of 2013 caused deaths among sheep and cattle. Flooding has proved an increasingly expensive problem for farmers by reducing pasture growth, hay and silage harvest. The threats from bluetongue virus and diagnoses of other diseases including severe rumen and liver fluke infections have increased concordant with winters becoming milder. These problems cannot now be ignored. By 2017, more than half of farmers in the UK were taking action to reduce greenhouse gas emissions by various actions including increased recycling of farm materials, better energy efficiency and more accurate addition of nitrogenous fertilisers. However, many still see control of gas emissions as of low priority when taking management decisions.

Despite widespread cynicism, increasing numbers of farmers are exploiting the opportunities that the new conditions offer. There are now landscape changes, mostly on a small scale so far, that have followed from this diversification. Previously almost unknown crops grown in the UK in 2016 included tea, sunflowers, sweet potatoes, water melons and walnuts. More dramatically, vineyards are spreading

Orchards may be at risk from mild weather that limits fertility in the absence of sufficient winter chilling.

Climate Change and British Wildlife

once again across the land. Between 1989 and 2013 the area devoted to vine growth increased more than twofold, though it was still only about 2,000 hectares in 2016. An increasingly benevolent climate means that English wines, at least the whites, are no longer a bad joke.

Maize is another crop very much on the increase in England. Since 1970 the land used for production of this forage plant has increased more than 150-fold, to around 160,000 hectares by 2012. Originally a subtropical organism, maize is on the edge of its climate envelope in the UK but is faring ever better as the climate warms. Unfortunately, it is one of the most unsustainable and environmentally damaging crops possible to grow in the UK. It is cropped late in the season by heavy forage harvesters, compacting the soil and leading to rapid water run-off. Somewhat ironically, maize fields are appearing on the Somerset Levels, one of the most flood-prone districts in the country. And there is more irony. Maize is a favourite food of Badgers, and its increased availability is probably one of the factors underpinning population increases in that animal, the self-same beast that livestock farmers want to cull.

It seems, then, that farmers are increasingly accepting of the issues raised by climate change and beginning to act on them, usually (according to surveys) motivated primarily by the prospect of financial benefits. Of course we need farmers to prosper, but it will be increasingly important that their actions take full account of their impacts on climate change. Growing maize may make money, but it is not a good example of what farming needs to do.

Gardening

Another outdoor activity for which the UK is especially renowned is gardening. The number of enthusiasts for this hobby is breathtaking: some 27 million people, 40 per cent of the population, are reckoned to engage in it to some extent. Here is another group well placed to notice effects of climate change, in this case on their garden flora. The behaviour of gardeners is of considerable significance, because the value of gardens as a wildlife resource has long been recognised. When suitably managed they can provide more wildlife-friendly habitat than much of the intensively farmed landscape beyond them. The total land area covered by gardens in the UK is impressive, amounting to more than 400,000 hectares (an area larger than Suffolk) with some three million ponds and perhaps 30 million trees.

OPPOSITE PAGE TOP:
Grape vines in Surrey. Vineyards are increasingly viable options for British farmers as growing seasons lengthen.

OPPOSITE PAGE BOTTOM:
Maize field in autumn, Yorkshire Wolds.

Lawn-mowing in autumn, an increasingly common pastime.

What gardeners do matters, so how do they feel about the impacts of climate change on their hobby? Some are responding by planting new species that are likely to thrive under warmer and drier conditions. This was evident at the 2017 Chelsea Flower Show, which featured Mediterranean plants such as White Henbane *Hyoscyamus albus* and Prickly Goldenfleece *Urospermum picroides*. Probably the most noticeable change for gardeners, though, has been the lengthening growing season and the need to mow lawns ever earlier and later in the year. This has not been a universally welcome development.

The Royal Horticultural Society (RHS) has not been slow to advise gardeners of both the problems and the opportunities offered by a changing climate, all in a report by Webster *et al.* (2017). Most of the respondents in recent RHS surveys agreed that climate change was under way, with increased unpredictability of the weather a foremost concern. One obvious response, namely the introduction of more non-native species better adapted to warmer summers or wetter winters, raises the risk of more invasive plants spreading in the countryside. This danger is recognised by the RHS, but it seems doubtful that its recommended solution is realistic. Gardeners are invited to reduce the possible expansion of invasive species by ensuring that cultivated plants remain in the garden. Just how seeds are to be deterred from travelling beyond hedges and fences is not explained. The more or less unrestricted ability of garden centres to propagate species from all over the world remains a major loophole in laws purporting to restrict the import of non-native plants and animals. Gardeners cannot be expected to take account of this risk when ever more novel species will be on offer, so perhaps this largely uncontrolled business should attract the attention of government regulators. That would not be a universally popular move, but it might prevent further disasters of the kind that foreign imports have caused on many previous occasions. Lady Anne Brewis, an amazingly knowledgeable botanist whose habitat included Gilbert White's old haunts around Selborne, was prone to rant in uncharacteristic fashion about her father's introduction of rhododendron to their estate in north Hampshire. Her anger was justified. This attractive but highly invasive shrub has run riot all over the UK to the detriment of many native species. There have been all too many other, similar, examples. Climate change must not provide an excuse to take more risks to our native flora by compensatory actions in our beautiful parks and gardens.

Naturalists and scientists

The responses of naturalists to climate change have been virtually unanimous in realising its consequences for British wildlife, as evidenced by many comments throughout this book. Robert Marsham identified a link between phenology and climate over 200 years ago, a time when climate change was happening as the Little Ice Age began to peter out, but in those days climate trends went largely unrecognised. The revival of phenology as a fascinating indicator of how climate is affecting British wildlife has been largely driven by Tim Sparks, a former professor in environmental change at Coventry University. The number of published papers he has authored, just a handful of which are cited in this text, is a testament to Tim's dedication to the subject. While at the Centre for Ecology and Hydrology (CEH), Tim helped persuade the Woodland Trust to establish the *Nature's Calendar* website in 2000. Judging by student feedback, he was one of Coventry's most popular lecturers, with responses including 'funny and passionate' (I can vouch for that), as well as 'if he doesn't like you, you're screwed' (about which I can hardly comment). Tim's contributions to the subject of this book have been groundbreaking and continue apace.

Tim Sparks, leading wildlife phenologist of recent times.

Variations in the distribution and abundance of plants and animals in the UK constitute the second major theme of climate change research. Professor Chris Thomas and his group at the University of York have been at the forefront of work in this area for some 20 years. As with Tim Sparks, a small sample of Chris's extensive publications

Climate Change and British Wildlife

Chris Thomas of the University of York, hunting for Orange Tip butterflies on Lady's Smock.

on the subject can be found in this book. Although specialising in insects, especially butterflies, Chris and his team have brought together results from many groups of organisms and demonstrated widespread trends for distributional ranges to shift northwards. His studies have extended beyond British shores, and in recognition of his major contribution to understanding the ecological impacts of climate change he was made a Fellow of the Royal Society in 2012. Chris errs on the optimistic side when contemplating the flow of events in the UK, emphasising the positive aspects of new species arrivals as the climate changes. At the same time he has recognised that it is not good news for all species and that elsewhere in the world climate change is adding to extinction risks. Chris's work is at the forefront of figuring out how British wildlife is responding to climate change and how conservation measures can be developed to accommodate these responses. In his own words:

It was not at all clear how species would respond to climate change back in the 1990s. We 'knew' that many other pressures were affecting our wildlife, such as the quantity and quality of habitat in a landscape, the extent to which the habitat was fragmented by human activities, and whether a species was persecuted or poisoned by pesticides and pollutants. These seemed far more important than the annual warming trend, which was dwarfed by yearly, seasonal and even daily fluctuations in temperature. But, eventually, curiosity got the better of us and we started counting dots (10km grid squares where the species was present) on distribution maps to define the northern boundaries of species at different times. It was immediately obvious that the Comma, Speckled Wood, Orange Tip, Small Skipper and several other species of butterflies had all extended their geographic ranges beyond the distributions that they had in the 1970s. Well, why stop at butterflies? I then scrutinised the maps that had been published by the British Trust for Ornithology in two British bird atlases, one with records dating from 1968–1972 and the other from 1988–1991. Again, the northernmost dots where breeding was taking place had, on average, moved further north by the time of the second Atlas period. The next issue was whether the same changes were taking place in other groups. The answer was 'yes'. Most other insect groups had indeed moved northwards. If the current rate of change keeps up, it will suffice to re-arrange every biological community on earth within a couple of centuries. Fundamental biological change is taking place, yet we hardly notice it!

As time goes by

James Pearce-Higgins is Director of Science at the BTO and leads climate change research there. Once again his publications permeate this book and, among other things, James has led many studies predicting future trends in bird population as climate change progresses. In his own words:

James Pearce-Higgins, Director of Science at the BTO.

> *In many ways, the study of climate change impacts is more challenging than the study of other environmental pressures or changes. By its very nature, climate change is only apparent over a timescale of multiple decades, which means that to document impacts of climate change, some form of long-term data is required. At its simplest, this can involve repeating past surveys or collections of data to look at change between two time periods. Time series of continuous long-term data can be more powerful, enabling not just the trend or change through time to be described, but variation in population abundance between years to be linked to annual variation in climatic parameters. The most robust studies are those which are not just correlative, but where there is more detailed mechanistic or ecological understanding that closely links the observed changes to climate change.*
>
> *One of the most secure ways of achieving the necessary large-scale and long-term data collection is through citizen science. For example, the UK Breeding Bird Survey run by the BTO supports 2,800 volunteers to survey over 3,800 randomly distributed 1-kilometre squares across the country to provide population trends of 111 bird species. More broadly, annual monitoring by over 8,500 volunteers produces trends for 250 bird, 56 butterfly and 20 mammal species across the UK. Although citizen science is not free, if done well it can provide a large amount of incredibly useful long-term data. In places where such monitoring does not exist, there is an urgent need for long-term monitoring to be established, in order to document the potential impacts of climate change into the future.*

Of course many other scientists have contributed to our understanding of what the climate is doing to wildlife in the UK and beyond. The commitment of this community, as well as its growth in size and influence, has been instrumental in persuading policy makers that the issues in our countryside arising from climate change cannot be ignored.

The culture of climate change

For a subject so newly arrived in the public consciousness, the response of artists and poets to climate change has been remarkably rapid. Despite its ominous overtones, the concern has caught the imagination of some very talented people able to convey its importance to an audience including those not following the vagaries of climate change science.

Jill Pelto has been in the vanguard of painters applying their skills to climate change issues, and there are some wonderful images out there, of which the one depicted below is a personal favourite. 'Landscape of change' brings together melting glaciers, rising sea levels and burning forests.

'Landscape of change': artwork by Jill Pelto.

Charlotte Sullivan's winning picture from Paint for the Planet, a children's art exhibition and auction to raise money for UNICEF's emergency relief for children affected by climate-related disasters.

Especially impressive has been the impact of climate change on artwork by children. Charlotte Sullivan was one of many youngsters contributing to an international art festival in New York. Just 11 years old when she created her masterpiece, it is surely a cause for optimism that she and others of her generation are already concerned about an issue that will bear heavily upon them.

It is fascinating to recall that some paintings of earlier times depicted what we now recognise as signatures of past climate change. Perhaps the best known are images of activities on a frozen River Thames several centuries ago, as illustrated on p. 12. There are, however, more recent examples that have also been interpreted as telling us something about climate history. In an imaginative synthesis of art and science, Zerefos *et al.* (2007) analysed the red–green colour ratios in sunsets depicted by famous painters between AD 1500 and 1900. Comparisons were made between pictures categorised as showing 'non-volcanic' or 'volcanic' sunsets, the latter being created within three years of a major volcanic eruption. Differences in the colour ratios were significant and, in the case of volcanic sunsets, were as predicted from the spectra of light scattering observed after recent eruptions. The sky in Turner's 'Fighting *Temeraire*', painted in 1838 shortly after a major eruption of Mount Coseguina in Nicaragua, does indeed look very different from a Petworth Park image created ten years earlier when there had been no recent volcanic explosions. Volcanic dust has a cooling effect and thus precipitates short-term climate change, as happened after the Mount Pinatubo eruption in 1991.

LEFT: Turner's 'Fighting *Temeraire*', painted in 1839 when volcanic activity was high.

RIGHT: Turner's 'Petworth Lake', painted around 1829 at a time of low volcanic activity.

Poets, too, have become engaged with climate change, and have created some moving work on the subject. Rachel McCarthy, a highly accomplished climate scientist, has become renowned for poems with a scientific bent. 'Survey north of 60 degrees' epitomises a deep concern for Arctic lands and the damage that a warming planet is wreaking upon them:

We're here to cast off names –
Viking, Fair Isle, Faeroes,
pronounce drowned coves, remap the coast.
I'm troubled – not at the cliff's seaward shiver
or the guillemots' black beaks scissored and shrieking –
but the wind singing
one long low note
its worm-burrow to the heart of the Arctic.

Late, in your hotel room, we nip at a bottle of Absolut,
talk of tongues of ice repealing themselves:
Novaya Zemlya, Svalbard, Barents.
I don't mention the wind tunnelling me
like the wisteria that arched the path
from the park to my childhood home
where I'd sneak a smoke
to inhale the boy I thought I loved
before I knew what love was;

snow-quiet, might, obliterative,
to be able to sit at the world's end
and say little of it.

Rachel McCarthy: poet, climate scientist, essayist and broadcaster.

Further afield, the dangers of climate change to people as well as wildlife are even more imminent. The Marshall Islands, a group of beautiful Pacific atolls, are under threat from rising sea levels and may soon disappear altogether. In 'Utilomar', Kathy Jetñil-Kijiner makes a passionate plea to rescue her homeland:

> *In the Marshall Islands I teach Pacific Literature*
> *Together we read the stories our ancestors told around coconut husk fire*
>
> *So what are the legends*
> *we tell ourselves today?*
> *What songs are we throwing into the fire . . . what*
> *are we burning?*
> *And will future generations*
> *recite these stories by heart, hand*
> *over chest?*
>
> *Maybe*
> *In one legend*
> *It'll start by saying*
>
> *in the beginning*
> *was water*
>
> *water from the sea that flooded our homes our land and now*
> *our only underground reservoir*
> *what we call a fresh water lens*
> *shaped like the front of an eyeball, nestled deep in our coral*
> *feeding on rainwater it watches us, burning and angry it is*
> *vindictive*
> *it poisons us*
> *with salt*
> *leaving us dry*
> *and thirsty*

Climate Change and British Wildlife

Al Gore at the Paris climate change summit in December 2015.

It doesn't stop there. Films about climate change include Al Gore's *An Inconvenient Truth* and the recent follow-up, *An Inconvenient Sequel: Truth to Power*, not to mention a flurry of science fiction contributions that have generally received three-star reviews or worse. And there are popular books, some serious but many fictional, and enough now to warrant their own genre with a 'cli fi' label.

On a lighter note, cartoonists too have captured the spirit of the climate change debate. Examples permeate social media and the web, but I am especially fond of the one shown here. Humorous insights into the mindsets of climate change sceptics are, for me, a powerful way of highlighting the absurdity of evidence-free arguments. The art world has more to offer in the climate change debate than is widely realised.

'Climate Summit' cartoon, from the *New York Times*.

Non-government Organisations (NGOs)

The best-known NGOs involved with wildlife conservation have unanimously campaigned for measures to slow down and eventually stop climate change. Friends of the Earth, Greenpeace, the RSPB, the Wildlife Trusts and the Worldwide Fund for Nature (WWF) have been among the most prominent advocates of remedial measures, including lobbying members of parliament and government ministers. Sadly it has largely been a dispiriting period for those with a green agenda, because even climate change, surely one of the most pressing issues threatening future generations, is scarcely on the political horizon most of the time. Wildlife is pretty much off it altogether despite the best efforts of the NGOs and some high-profile individuals such as David Attenborough and George Monbiot. It seems the best that can be done is to keep on plugging away, and there are signs that some of the great and powerful are beginning to pay the subject (climate change, sadly not so much wildlife) the attention it deserves.

Business leaders and politicians

Businesses have become increasingly aware of both the dangers and the opportunities resulting from climate change, especially after the 2006 report by prominent economist Nicholas Stern. Stern is that rarest of beasts, an economist with an understanding of how economic activity affects the environment. Unsurprisingly there was a barrage of criticism from more typical members of the economics fraternity who prefer to inhabit a cocoon isolated from the real (i.e. biological) world. Ten years on, Stern was unrepentant, even commenting that his report underestimated the risks. Fortunately the results were widely accepted by many politicians and activists, but other important organisations, such as the Institute of Directors, still seem cautious about giving climate change issues a high priority. While many of its members support taking mitigation measures, a sizeable number also support fracking and continue to advocate expanding UK airports and flying capacity.

The Confederation of British Industry is supportive, at least in principle, of methods to combat climate change. Indeed, its director John Cridland has criticised government policy reversals such as backing away from subsidising renewable energy. And Governor of the Bank of England Mark Carney, another unlikely champion,

Nicholas Stern presenting conclusions from his review in October 2006.

has promoted awareness of climate issues as risks to the financial sector – and has been damned for doing so by other economists and politicians. It seems that realisation of how serious climate change could become is sinking home in the business sector, which hopefully bodes well for the future.

Climate change arrived on the political agenda relatively early in the UK, starting with Prime Minister Margaret Thatcher's recognition of its significance in the late 1980s. Since then there has been widespread acceptance across the political spectrum of the evidence for climate change and its implications for the future of the global environment. There have been regular updates, ever more compelling, from the Intergovernmental Panel on Climate Change (IPCC); a UK Climate Change Programme was established in 2000; and the nation's Climate Change Act took effect in 2008. None of this was specifically directed at impacts on wildlife, the focus being on reducing carbon dioxide emissions. However, attitudes towards climate change policies have never been unanimous in parliament, and several Chancellors have resisted responses that would curb greenhouse gas emissions.

Economists seem particularly ignorant of, and hostile to, influences outside the narrow remit within which they operate. David Cameron's 'greenest government ever' quickly metamorphosed into one keener on denigrating 'green crap'. Since then the attitudes of government ministers towards mitigation measures against climate change have been mixed and sometimes personally inconsistent. What are we to make of this ambivalence? Sadly, the UK has yet to produce a politician taking the climate change agenda as seriously as one-time US Vice President Al Gore, whose commitment brought the environment to the top of the political agenda and earned him a Nobel Peace Prize. Will Britain, I wonder, ever have a leader of that calibre?

Climate change will remain on the political agenda for the foreseeable future, but an overriding concern for British wildlife is what will happen after the UK leaves the European Union. While there will be opportunities for improving wildlife conservation, many of the politicians best known for their desire to cut red and green tape (the 'green crap' brigade) are still alive and kicking. It therefore remains to be seen whether dominant personalities in parliament will rise to the daunting challenge of improving environmental protection, with policies supporting green energy levies and the like, or fall back on populist agendas in a race to the bottom. It is down to all of us to keep lobbying for the first option.

Then and now

In the autumn of our years it is tempting to reflect on changes witnessed over a lifetime. A lot has happened to British wildlife over the past half-century or so. Pond pollution has increased while rivers have become cleaner. Wildflower meadows are almost a thing of the past but reedbeds are on the increase. Songbirds have declined while raptors have resurged. Many butterflies have become rare while Polecats and Pine Martins *Martes martes* have bounced back, and Wild Boar *Sus scrofa* and Beavers *Castor fiber* have returned after centuries of absence. Further possible additions to our fauna are making the news, notably a controversial discussion about reintroducing Lynx to Northumberland.

People have heard about or noticed these changes. But what about memories of how the climate used to be? Inevitably these are more vague, and often coloured by the most dramatic events, those hot summers and bitter winters. Only one trend really stands out as a common experience by almost everyone asked about their perceptions of climate change: winters are not what they were. Childhood recollections from the 1950s include sledging and snowball fights almost every year, while today's winters tend to be damp and bleak, less fun and with fewer school closures for kids today.

Rarely have people noticed a clear connection between climate change and wildlife, for the simple reason that its effects have been smaller, and less obvious, than other changes in the countryside. More cheeringly, people are nevertheless recognising the need to conserve wildlife and joining effective organisations dedicated to that end, such as the RSPB, in ever-increasing numbers. This is what really matters, whether or not reasons for the need are fully understood.

UK residents seem to focus particularly on wet-weather-related events when it comes to forming their beliefs about climate change.

new areas quickly, with a short life-cycle, self-fertilisation, wind-based dispersal and rapid germination. Modellers have attempted to distinguish whether climate change, habitat modification or both have contributed to its range expansion, as described by D'Andrea *et al.* (2009). Information about the distribution of this plant across Europe, going back as far as the 1820s, was collected from an impressive range of sources including records from herbaria. Climate envelopes were devised to encompass Prickly Lettuce distributions during five different time periods (1901–1920, 1921–1940, 1941–1960, 1961–1980 and 1981–2000) for which adequate climate data were available across the continent. This allowed the testing of climate envelope models from each time-slice to be tested against each other, a validation that worked out well.

The spread of Prickly Lettuce has been only partly explained by climate-based models.

Key features of Prickly Lettuce's climate requirements turned out to be temperatures in the spring of greater than 5°C, temperatures in the summer between 7°C and 15°C, 2–10 months of temperatures greater than 10°C, less than 300mm of rain in winter, more than 300mm of rain in spring and summer, and less than 200mm of rain in autumn. Although the model incorporating these variables passed muster, we have no insights about how they influence the physiology of the plant. Why does a drought-tolerant organism need so much rain in spring and summer? Colonisation of much of Europe, from the plant's Mediterranean origins, was progressive through much of the 19th and 20th centuries. This in itself is a telling observation because, although the spread of Prickly Lettuce followed quite accurately the widening arc of its climate envelope, climate change was not very marked during this period of rapid range increase. However, what has happened since 1990 is illuminating. The climate envelope extended more rapidly than before, as might be expected, but the distribution of Prickly Lettuce has not followed suit. Western Scotland and much of Sweden became climatically adequate in recent decades but colonisation has not yet followed. There are at least two possible reasons for this discrepancy. Maybe it is just a lag, and the plant will appear in these places in due course. Alternatively, increasing levels of human disturbance creating readily colonised open ground may be more important than climate in dictating the distribution of Prickly Lettuce. The modelling study, interesting though it is, did not resolve that question. What it did do is demonstrate that, even when confined to only climatic factors, the envelope model for this species required a complex combination of temperatures and rainfall at different times of year. Nothing as simple as a single isotherm would suffice.

Another target for modelling studies is concern about the future prospects of arctic-alpine plants of the high mountains, for some of which the impact of climate warming is already causing decreases in abundance. Snow-bed Willow is one of the species at risk (see Chapter 6), having declined by around 50 per cent in the Scottish Highlands since the mid-20th century. It grows in exposed fell-fields, screes and snow-beds and requires habitats with some disturbance to persist. Despite its name, Snow-bed Willow needs to be free of snow for at least three months to flower and fruit; seeds are dispersed by the wind, but successful reproduction by this method is uncommon and most plants spread by vegetative growth. Modelling the climate envelope of this plant used a combination of current distribution information in Europe together with fossil records from as far back as the last glacial maximum about 20,000 years ago, as investigated by Alsos *et al.* (2009). Leaves of Snow-bed Willow are distinctive and fossilise well, providing a more accurate indicator of past distribution than pollen, which might travel some distance from its parent. Contemporary and inferred historical climate data were then invoked to create climate envelope models. Those based solely on the British distribution were tested for accuracy using random partitions of the data, and by determining how well they accounted for the plant's wider distribution across Europe. Just three climatic variables sufficed to produce an accurate envelope, namely maximum summer temperature, growing degree days (see p. 142) and mean summer water balance. The southern distribution limit of Snow-bed Willow, for example, closely follows maximum summer temperature isotherms of between 23°C and 25°C in the UK. There was a good correspondence between the modelled past distribution of the willow and areas with glacial fossil records, giving extra confidence to predictions about the future. Of course the central point of this exercise was to predict how well Snow-bed Willow will fare under warmer climate scenarios. Even

What the future may hold

Snow-bed Willow (Dwarf Willow) is predicted to decline in south European mountains but to persist in the UK.

when based on the least severe climate change expectations, a 46 per cent reduction in suitable areas for Snow-bed Willow across its entire range is anticipated by 2080. However, the major losses are predicted to occur in mainland Europe, and southerly outposts including the Pyrenees and Carpathians will probably become much less suitable than they are today. Better news for the UK is that this plant is still expected to survive in Scotland.

Climate envelope models are a useful method for predicting future distributions but they have limitations beyond the obvious one of ignoring other ecological factors. In both of the examples described above, a series of critical climate variables was identified, but exactly how they affected the plants remained unknown. A more comprehensive model would take account of exactly what the climate was doing to influence the plant response, and if successful should generate a more refined prediction of what will happen as climate change proceeds. To do this, however, requires detailed information about plant life-history traits including growth rates, time taken to reach sexual maturity, fecundity and possible trade-offs between these features. Taking longer to mature, for example, might lead to high fecundity but carry a greater risk of premature death than flowering at a younger age.

This kind of information is not available for many species but was obtained by Williams *et al.* (2015) for Lady Orchids in Belgium by monitoring two populations continuously between 2003 and 2013. These orchids are perennials, sustained by underground tubers that are replaced annually by new ones from which the plant regenerates in the following year. Leaves emerge in early February and are fully developed in May, together with the flowering stalk. Flowers have no nectar, and fruit-set is typically achieved by less than 20 per cent of the flowers that appear. Seed capsules ripen by the end of June and are dispersed in August. From the middle of that month onwards, nothing green is visible above ground. Flowering in Lady Orchids carries a trade-off because it comes at the cost of reduced vegetative growth.

A complex modelling procedure was developed, firstly to integrate all the life-history details of Lady Orchids with climate data and assess how each trait responded to temperature and rainfall variation; secondly to figure out how these responses affected the fitness of individual plants, and thereby to estimate the optimum sizes of flowering plants that would be selected for over time; and finally to assess how, taken together, climate change will affect the future prospects of Lady Orchids. The results were complicated but fascinating. Climatic influences were strongest on traits related to reproduction. The probability of flowering in any particular year increased following a wet spring in the previous year, and a dry winter in the current one. This effect was most marked in the larger plants. New recruits to the orchid population also responded to this combination of climate features by vigorous growth, and were more than two-and-a-half times larger in years following previous wet springs than in those following dry springs. Warm years had a separate effect, reducing the number of flowers produced per plant. Plant growth, on the other hand, was highest in years following warm winters and wet summers. Flowering orchids grew more slowly than non-flowering ones, but the vegetative plants were more strongly inhibited by dry summers, reducing the difference between them and those that were flowering.

Evidently the life of a Lady Orchid depends on a range of different climatic features, each of which influences different aspects of the plant's physiology. How does knowing that help to predict the consequences of climate change? The next stage of the model integrated the demographic and climate information to estimate how the size at which an orchid flowers under various conditions must have evolved over time. Increasing the frequency of wet and warm conditions selects for flowering at large plant sizes, which in turn imposes a delay on the age of flowering. By contrast, an increased frequency of dry seasons, colder winters or cold years decreases the size of flowering plants and thus raises the probability of smaller plants reproducing.

Finally, the model assessed the meaning of all this for future responses of Lady Orchids to various possible types of climate change. In a nutshell, the model predictions were optimistic. Whether due to genetic variation or individual plasticity, the flexible responses seen in this study should allow survival and successful reproduction of Lady Orchids under widely different climatic conditions. Taking account of all the different factors at work on various aspects of the orchid's life-cycle, it therefore looks as if this pretty plant should prove resilient to quite large changes in climate.

Climate envelope modelling will probably remain the most popular approach to predicting the impact of climate change on plants, because for most species there is insufficient information about their demography to do much else. It will be interesting to see, in the fullness of time, the degree to which more sophisticated approaches of the kind employed with Lady Orchids outperform simpler envelope models – if, indeed, they do.

Invertebrate models

The Green Hawker dragonfly *Aeshna viridis* is an interesting subject for distribution modelling for two reasons. It is not native to the UK but occurs on the near continent and looks like another possible invader in the near future. There is at least one record of it from the north-east coast of England and, as discussed in Chapter 3, several other members of this highly mobile group of insects have successfully colonised the UK in recent years. Will Green Hawkers follow suit? That is an intriguing conjecture, but something else about this species adds an extra challenge to modellers. It has, very unusually among dragonflies, a strong preference for laying its eggs on a particular aquatic plant. Most dragonflies and damselflies lay their eggs on any suitable-looking vegetation or just drop them haphazardly in the water. Not so the Green Hawker. Females of this fine animal seek out Water-soldier *Stratiotes aloides* on which to oviposit. Although not an absolute link, the local distributions of plant and insect are strongly

Green Hawker dragonfly laying eggs on Water-soldier.

matched. Water-soldier will be familiar to many pond dippers, with its star-shaped, floating clusters of sharp-toothed leaves more than capable of scratching an unwary hand. These leaves are believed to provide protection for the larvae of Green Hawkers against fish predation, and are presumably the reason for their selection as egg deposition sites. Water-soldier is native to the UK and is locally abundant, especially in the east of England where, perhaps, a new dragonfly is most likely to arrive.

This binary interdependence provided modellers with an opportunity to test how climate envelopes perform when more than just the climate for one species needs consideration, and was investigated by Jaeschke *et al.* (2012). It was still a relatively simple case, adding one extra environmental factor, the distribution of a host plant, to model the future distribution of an insect. Various possible scenarios were compared with climate envelopes for the dragonfly alone. In one scenario, dragonfly and host plant were modelled separately to show areas of future distribution overlap based on their individual climate envelopes. Another model explicitly incorporated the need for Water-soldier as an extra variable, on top of climate, in determining the dragonfly's future distribution. It turned out that all the models predicted the invasion of East Anglia by Green Hawkers by between 2021 and

2050 under a 'medium' expected rate of climate change. Elsewhere, the extent of the dragonfly's range expansion across northern Europe was reduced by 10–20 per cent by models that took the interaction with Water-soldier into account. We shall see, but a striking result of this study is the realisation that even adding just one extra environmental variable to a climate envelope can substantially modify model predictions of range changes.

Another limitation of climate envelope models as predictors of future distributions is that they rarely take account of dispersal ability. There is often a tacit assumption that if a 'field of dreams' exists, where suitable habitat has become available, then sooner or later it will be occupied. This is manifestly untrue for sessile animals with specialised habitat requirements. If there were sand dunes in Birmingham, Natterjack Toads would never get there despite the MONARCH model's projections because the intervening habitats are unsuitable for the amphibian.

Even for mobile species it is naive to assume rapid or inevitable colonisation of a newly available habitat patch. Modelling by Jaeschke et al. (2013) has addressed the question of how mobility influences dispersal and colonisation by European dragonflies and damselflies, with interesting results. For six species, only one of which currently inhabits the UK (the Southern Damselfly *Coenagrion mercuriale*), there was empirical evidence about how far the adults actually fly as well as the times taken for larval development. Dragonfly and damselfly nymphs typically spend between one and three years in the water prior to metamorphosis, and for serial dispersal from one 'stepping stone' pond to another this development period inevitably generates a lag before the next round of dispersal can even begin. Southern Damselflies in the UK spend two years as larvae, and adults can disperse at about 1km per generation. Models compared the consequences of three different dispersal options for the six species in the study: no dispersal, in which only declines are possible; full, essentially unlimited dispersal; and species-specific dispersal. In all cases the ambition was to predict range changes by 2035 under 'moderate' climate change expectations. Under the first model, lacking dispersal, all six species were predicted to lose half of their existing habitat spaces. According to either of the other dispersal models, two species still face the prospect of 30 per cent habitat loss, while another is set to expand its range. The remaining three species, including the Southern Damselfly, are expected to increase their overall ranges if they enjoy

Southern Damselflies are predicted to extend their range northwards in England, habitat permitting.

'full dispersal' or contract if the species-specific dispersal rates turn out to be more realistic. With species-specific dispersal, Southern Damselflies are projected to disappear from most of central Europe and Iberia because of increased aridity in their existing habitats, with a range contraction towards France and northern Spain. On the other hand, this species is predicted to increase in abundance and extend its northern distribution limit in the UK. However, this assumes that there is sufficient of its specialised habitat, namely small calcareous streams in open country, to constitute its 'field of dreams'. This is unlikely, again highlighting the limitation of models that take no account of habitat availability.

Despite this caveat, the qualitative differences in future prospects generated by models suggesting either range expansion or contraction, depending on dispersal behaviour, are a profound result with widespread implications for climate envelope modelling. The problem is universal, because every species will have limited dispersal ability relative to the distance it will need to travel to find future climate and habitat space. This presumably means that many climate envelope models that ignore dispersal will be overly optimistic predictors of distribution increases.

What the future may hold

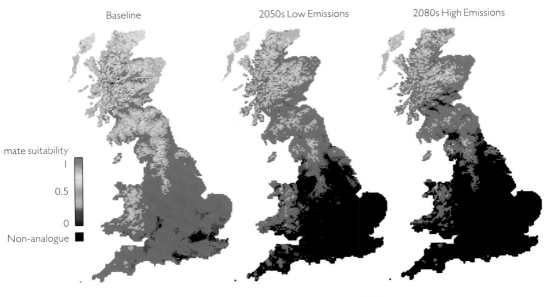

Projections to show suitable climate space for the arctic-alpine montane lichen *Flavocetraria nivalis* (pictured left) during a baseline period (1960–2000), and for the 2050s and 2080s under low and high greenhouse gas emissions scenarios. Black 'non-analogue' regions are unsuitable habitat outside the remit of the model. Courtesy of Chris Ellis.

Pyrenula macrospora looks set to extend its range dramatically into the Midlands, southern England and as far as the coast of East Anglia.

With high-altitude montane lichens apparently at particular risk from increased warmth, different models explored what might happen to them by including the way plant and lichen communities are moderated by wind speed in exposed locations. In the Cairngorms, arctic-alpine ground-layer lichens share a demanding, wind-clipped heathland habitat at altitudes between 700 and 950m, below the snow line but above the upper limits of dense macrophyte stands. Models including wind speed as a variable suggested that its effects on lichens in this habitat are indirect and work by controlling the extent, and by suppressing the height, of ericaceous heath. Bare ground is made available for lichens by limiting the higher plant competitor. Looking to the future under a high-gas-emission forecast, increased temperature alone would favour competitor plants and decrease substantially the abundance of ground-layer lichens on mountain slopes below 950m by the 2050s. Decreasing wind speeds would make the decline even worse. However, increasing wind speeds would counter the damaging effect of higher temperatures, and an average speed increase of 20 per cent would remove it completely.

As the modellers confess, predictions based on climate envelopes are particularly difficult for lichens. Those models incorporating wind speed did not include precipitation and, as described in Chapter 2, the effects of inclemency in high mountains alter according to whether it is manifest as rain or snow. Future wind conditions are among the most uncertain of the factors included by climate change modellers, though as a general increase in storminess is on the cards it is not unreasonable to expect some degree of average wind-speed increase. With so many imponderables, it is hard to judge what will happen to the British lichen flora, but the modelling community has raised awareness of what we should look out for. The indications are that montane species, at least, share with arctic-alpine plants a very uncertain future.

North Sea models

Peering into the future of a species is one thing; planning for the future of an entire ecosystem is quite another, but that is what has been attempted for the North Sea. A primary motivation for such an ambitious exercise has been commercial rather than ecological, which is hardly surprising considering the importance of the fishery to countries surrounding this tempestuous waterway. Models of the North Sea communities come in two forms. Some attempt to explain the current situation and how it has come about, while others go further and attempt to predict what will happen in the future.

One approach towards explaining the current run of events is to investigate regional variations in the North Sea environment and link them to biological consequences. According to one model, the influences of temperature and salinity on recruitment of fishes including Atlantic Cod, European Plaice *Pleuronectes platessa*, Herring,

Dover Sole and European Sprat *Sprattus sprattus* vary according to geographical location, as indicated by Akimova *et al.* (2016). Warming of the north-westerly region around Scotland is associated with poor Cod recruitment, whereas temperatures in the south-east, in the German Bight, are problematic for Plaice. Sprats fare best in regions of highest salinity, especially along Britain's eastern coast.

This kind of analysis makes an important point: the North Sea cannot be considered a uniform environment, given its wide ranges of latitude and depth. Models of possible impacts of acidification tell a similar story, varying across the region and potentially having local effects on fish abundance as great as those of increased temperature. However, modelling suggests that wide-ranging changes in pelagic (mid- or surface-water) fish populations between 1965 and 2012 were primarily driven by temperature increases. There has been a marked shift from cold-water assemblages towards subtropical ones since the 1990s, with declines in Sprat and Herring concomitant with increases in Atlantic Horse Mackerel *Trachurus trachurus*, Atlantic Mackerel *Scomber scombrus*, Sardines and Anchovies. In all cases, sea surface temperatures were the most important model variable explaining the distributional changes.

There have also been attempts to explain what has happened to one particular species of great ecological importance, the Lesser Sand Eel. Between 1973 and 2006 growth rates of young sand

Computer models of the future North Sea anticipate increases in fishes such as Atlantic Horse Mackerel.

eels dropped by 22 per cent, a decline that accelerated after 2002. Models developed by Frederiksen *et al.* (2011) suggest that key factors behind the decline include changes in the seasonal distributions of copepods as well as competition with planktivorous fishes, but once again these ecological effects are ultimately attributed to increases in water temperature.

The significance of sand eels to the fate of North Sea Kittiwakes has become widely accepted, and another level of modelling attempted to disentangle the effects of temperature and commercial sand eel fisheries on Kittiwake breeding success between 1986 and 2002 (Frederiksen *et al.*, 2004). Both factors seem to have contributed to the Kittiwake's predicament. Fishery closures coincided with improved breeding success of the birds, while warm winters had the opposite effect, almost certainly a result of lower sand eel recruitment. An important conclusion of the model is that exploitative fisheries, not just climate change, can have seriously damaging effects higher up the food chain. Aside from the extra stress on seabirds caused inadvertently by humans harvesting their main food resource, just about every model of North Sea ecology points the finger at temperature increase as driving recent community changes. Even models in which the North Atlantic Oscillation (NAO) accounts for most variation in the recent breeding success of North Atlantic seabirds home in on its effects on sea surface temperature as the factor that really matters.

What, then, of the future for North Sea fish communities? Modellers have not been daunted by the enormous challenge posed by complex ecosystems. The most straightforward approach has entailed computing climate envelopes for multiple species individually and assessing how the fish distributions, and thus the community, will change over time. Taking this line together with a high emissions scenario, three different modelling methods and two climate data sets were applied by Jones *et al.* (2013) to assess the future distributions of ten commercially important and seven threatened species of North Sea fishes, assuming unlimited dispersal. Using multiple models was an attempt to compensate for the uncertainties inherent in any one of them, with a view to arriving at a consensus.

Although model predictions varied somewhat, a common pattern did emerge. All the species studied are expected to shift their range centres northwards by an average of 27km per decade. Range areas are not predicted to change significantly, and neither are range overlaps. The overall impression is of a relatively straightforward and cooperative northward shift for the entire fish community. However, this bland conclusion hides some fairly marked differences among the models with respect to individual species. Angelsharks *Squatina squatina* are predicted to experience a small range contraction by two of the models, while the Common Skate *Dipturus batis* will lose 11.6 per cent of its current suitable habitat based on one model combination. However, the critically endangered Skate and the Sand Ray *Leucoraja circularis* might also experience net gains in suitable habitat, of over 40 per cent and 10 per cent respectively, if other model combinations turn out to be the most accurate. We will have to wait and see whether one of the models eventually wins out and is the most reliable indicator of future times, or whether averaging across models adds the anticipated extra level of forecast dependability.

Not all North Sea modellers have been driven by interests in fishes. One study, by Wethay *et al.* (2011), looked at the impact of past extreme weather events, specifically severe winters, on distributional adjustments of marine invertebrates and used these past changes to predict future ones under a warming climate based on moderate emissions. Unusually cold British winters in 1962/63 and 2009/10 had unexpectedly profound, albeit temporary effects on two 'northern' species, the barnacle *Semibalanus balanoides* and the Lugworm *Arenicola marina*. Both of these animals recruited well in years subsequent

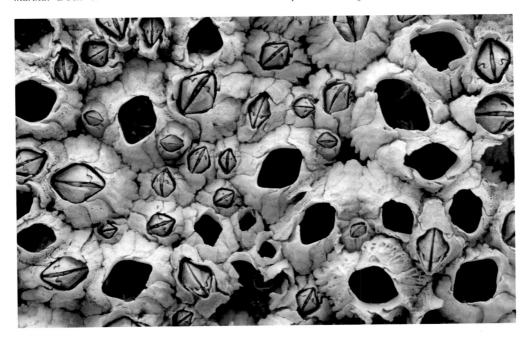

Cold-water barnacles, here at Kinloch Bay on the isle of Rum, are expected to retreat and be replaced by warm-water species.

to the cold spells, spreading southwards around France and even arriving as far south as Iberia. By contrast, two 'southern' barnacles, Montagu's Stellate Barnacle *Chthamalus montagui* and Poli's Stellate Barnacle suffered recruitment failure, though adults survived. Other 'southern' species, including a polychaete worm, a limpet and a mussel, remained more or less unmoved by the cold weather. The modelled effects of climate change based on these trends include, by 2090, the disappearance of water in the English Channel sufficiently cold to block the dispersal of southern species into the North Sea. If this actually happens and the climate envelope models for marine invertebrates are correct, community changes parallel with those predicted for fishes are clearly on the cards, although dispersal will likely be slower for the invertebrates than for the vertebrates.

Most modelling of the future North Sea ecosystem has, perhaps wisely, focused on limited sets of species. A more ambitious enterprise, based on moderate greenhouse gas emissions, attempted to describe the future distributions of 47 seafloor species by 2099 (Weinert *et al.*, 2016). The model assumed that during the 21st century the average temperature at the bottom of the North Sea would increase by between 0.15°C and 5.4°C depending on time of year and location, and salinity by 1.4 units (that is, to about 4 per cent above current levels). Climate envelopes, and therefore both present and future

The climate envelope for brittle stars will probably move north, possibly beyond British shores.

distributions, were modelled separately for each species. Among the 36 epifauna (animals living on the seabed), 50 per cent are expected to shift their distributions northwards by tens of kilometres. The brittle star *Ophiothrix fragilis* might manage more than 100km.

Some model predictions, such as those for hermit crabs, seem counterintuitive.

Curiously, though, the other 50 per cent are predicted to move similar distances southwards, with the hermit crab *Pagurus pridaeux* going for a record of around 100km. Of the 39 infauna (animals living within the seabed, such as various worms), 77 per cent will head north, but by smaller distances, on average, than the epifauna. The model also predicts longitudinal shifts, both east and west for species moving north, but mostly eastwards for those going south. Unlike the proposed future of most North Sea fishes, the outcome for the geographical ranges of benthic invertebrates does not look benign. Fifty-eight per cent of the epifauna and 72 per cent of the infauna are expected to lose ground, especially those moving north. In five extreme cases, including the brittle star, their North Sea habitat is predicted to eventually disappear altogether. Going south, the hermit crab could experience a range increase of perhaps 90 per cent. Overall, 60 per cent of species are expected to be net losers in the extent of their geographical ranges.

These predictions beg explanations. Firstly, why should animals moving north experience range contractions rather than expansions?

The suggestion is that the steep depth gradient across the North Sea, dropping off below 50m with a contour running roughly from Yorkshire to north Denmark, will serve as a barrier to northward range expansion because of sharp temperature transitions across it. The southern part of the North Sea is much shallower, and has warmed more than the northern section.

Predicted southward shifts of northern animals are more difficult to explain. Species moving in directions contrary to climate change expectations are not unprecedented, but the scale of such aberrations inferred from this model is unusually high. Anticipated longitudinal changes were often as great as latitudinal ones, again requiring justification. The gradient of the North Sea floor runs downwards from south-east to north-west, so it may be that animals moving south would find habitats to their liking in the German Bight, in keeping with the eastward movement predicted for them. They would, however, be venturing into a warmer habitat than the one they were leaving, a counterintuitive response for northern, predominantly cold-water organisms.

We can only wait and see whether this most ambitious of modelling exercises, with some curious expectations, comes to fruition. If it does there could be dramatic impacts at the community level. Some of the animals predicted to decline substantially include the brittle stars *Acrocnida brachiata* and *Amphiura filiformis* as well as the sea urchin Sea Potato *Echinocardium cordatum*. These species play important roles in disturbing soft sediments, redistributing organic matter and stimulating the recycling of nutrients. They may be replaced by southern species, but, if not, the future structure of the North Sea ecosystem could alter in an unpredictable fashion.

Of particular interest for wildlife conservation is the relationship between events in the North Sea and the viability of the UK's enormous cliff-nesting seabird colonies. One doom-laden model from Frederiksen *et al.* (2013) addressed this question by predicting how the habitat niche of *Calanus finmarchicus* will change during the 21st century, assuming moderate greenhouse gas emissions. As discussed in Chapter 7, there is plenty of evidence to show that the abundance of this copepod is critical to the breeding success of seabirds that rely heavily on Lesser Sand Eels to feed their chicks. The model bypasses intermediates in the food chain and simply links copepod profusion to seabird breeding success, based on surveys of Puffins, Guillemots and Kittiwakes on the Isle of May in Scotland

and on the island of Røst in northern Norway, over periods between 1980 and 2010.

The niche model for *Calanus finmarchicus* was primarily dictated by sea surface temperature, and, on this basis, its niche suitability has declined continuously off the east coast of Scotland since the 1960s. Post-2000 there has been a similar trend off southern Iceland, while the niche has remained stable around the Faroe and Lofoten islands, including Røst. However, environmental suitability for the copepod is predicted to decline strikingly all across the North Sea during the 21st century. Given continued climate warming, *Calanus finmarchicus* abundance sufficient to support optimum breeding success by Kittiwakes, Puffins or Guillemots is likely to be inadequate off the Scottish coast by the 2020s and around the Faroes by the 2060s. This prediction may even be optimistic if large-scale sand eel fisheries continue to exacerbate food shortages for the birds. On the other hand, alternative nutrition may reduce dependence on sand eels if sardines or anchovies can take their place. Unfortunately, it seems likely that even if the southern copepod *Calanus helgolandicus* replaces *Calanus finmarchicus*, it will not sustain high sand eel numbers because, as described in Chapter 7, it does not reach comparable abundance in the critical spring period. The overall conclusion of the model – maybe overly pessimistic – is that the large seabird colonies that astound and delight visitors to the UK's northern, craggy shores may one day be no more than a memory. One glimmer of hope is that Kittiwakes on the west coast are less dependent on sand eels so may fare rather better than those around the North Sea.

What can we make of the various attempts to predict how the North Sea ecosystem will look in the coming decades? Most models have confined themselves to subsets of the fauna, such as fish or invertebrate communities, and have based their projections on climate envelopes for each individual species. This is undoubtedly the most realistic approach in terms of keeping complexity within manageable limits, but it is inherently circumscribed because it cannot consider the system as a whole. Predictions of how fish distributions will change, ignoring different projections for the invertebrates upon which they feed, must be treated with great caution. Temperature increase is accepted as the main driver of future change, but inevitably there will be other factors of unpredictable significance, not least the impact of future fishing pressures. Modelling entire ecosystems, even in part, is a heroic enterprise, and it will be fascinating to see how model predictions pan out as the years go by.

Experiments

For those looking to the future of British wildlife, experiments provide an alternative approach. However, experiments, like computer models, share one important attribute: an assumption about how the climate will change and, specifically, how far temperatures are likely to increase. Temperature is usually the critical variable in these experiments, carried out at various scales, some inside laboratories, some in outdoor arenas and a few in the open field. Usually they focus on the responses of one or a few species under controlled conditions that attempt to exclude variables other than the one of primary interest. In going down this route we might usefully recall some remarks of the 19th-century scholar Ralph Waldo Emerson:

> *Science ... assumed to explain a reptile or mollusk, and isolated it – which is hunting for life in graveyards. Reptile or mollusk or man or angel only exists in system, in relation.*

Ecologists in particular among the scientific fraternity will surely recognise the difficulties of extrapolating from simplified experimental conditions to what actually goes on in the field. Nevertheless, there are questions about the consequences of climate change that are best investigated in the laboratory, and indeed in some cases this is the only possible route.

Plant experiments

Plants are good subjects for experimental work because many grow easily in greenhouses, and in the outside world they stay put wherever they take seed. As mentioned previously, the Snow-bed Willow is one of several arctic-alpine species considered to be at risk from climate warming, and it is already declining in the Scottish Highlands. This is a dwarf, perennial shrub that occurs naturally in exposed ridge and snow-bed habitats, but climate change is causing snow-bed sites to thaw earlier than in the past. Experimental transplantations of willows in the Swiss Alps by Sedlacek *et al.* (2015) were designed to test whether fitness differed in sites where snow melts relatively early, compared with others where it persists for longer. The moves were replicated at six sites, with multiple, reciprocal transplants between each pair of exposed ridges and snowfields, where snow melted a month later than on the ridges. Following transplantation, each willow was monitored for two years to record a range of traits thought to reflect fitness.

Leaves appeared earlier in the ridge habitats, but growth, once started, was quicker in the snow-bed sites and eventually produced larger leaves than were seen on the ridge plants. The willows showed plastic responses to the changes in their environment for these growth characteristics, with transplants behaving the same way as shrubs in the location to which they were moved. In contrast, flower production and stem number (a measure of vegetative growth) of transplants remained similar to those in the habitats of origin. It seems that Snow-bed Willows can modify their phenology in response to reduced snow cover but are unlikely to have habitat-specific genetic adaptations. In the long term, it looks as if reduction in leaf size and flowering together with longer development time may compromise the fitness of Snow-bed Willows under conditions of earlier snow-melt. Whether the consequences of physiological effects identified in this study eventually lead to plant death and explain ongoing population declines of the willow are connections yet to be made.

This kind of experiment, like the investigations of Sessile Oak bud-burst described in Chapter 2, is designed to find out whether responses to climate change involve both the ability of individuals to change their phenology (plasticity) and genetic variation among different individuals. The relative importance of nature and nurture as determinants of adaptation to climate warming is likely to receive increasing attention in future. For individual plants or animals, plastic responses are the only option, and for long-lived species plasticity may be pushed to or beyond its tolerable limits if climate change continues apace. The high rate of temperature increase also means that genetic responses, if possible at all, will mostly depend on existing variation. Waiting on new mutations will generally take too long to be of much help, although short-lived species with large population sizes and rapid generation turnovers may conceivably benefit from novel genetic adaptations.

Experiments are the only way to dissect the mechanisms underpinning climate responses, and hopefully they will lead to discoveries useful for conservation. The discovery of Ash plants resistant to dieback disease is a recent example of how genetic information can be used profitably, in this case to begin the fight-back against a pathogen threatening to devastate British woodlands. Perhaps it will also prove possible to identify and promulgate plants, even animals, able to survive future temperature hikes.

Invertebrate experiments

Insects are among the best animal subjects for experimental work, as evidenced by the huge contribution that studies with fruit flies have made towards the understanding of genetics. In the UK, the Mountain Ringlet *Erebia epiphron* is confined to high-altitude grassland sites in Scotland and the Lake District and, as a cold-adapted species, is potentially vulnerable to climate change. Experimental studies on the closely related Woodland Ringlet *Erebia medusa* in Germany suggest that this concern might be justified.

The Woodland Ringlet is a European butterfly adapted to a continental climate of warm summers and cold winters. Experiments by Stuhldreher *et al.* (2014) were designed to investigate how warmer winters could affect Woodland Ringlet populations by exposing larvae, the overwintering stage of its life-cycle, to three temperature regimes. The coldest treatment simulated winters at a site where the butterfly continues to thrive; an intermediate treatment mirrored conditions where the insect survives but has declined; and the warmest treatment corresponded to winter temperatures immediately south of its current range limit. By the beginning of March, almost all of the larvae reared in the warm treatment had started feeding, far in advance of those in the two cooler treatments. Warm-treatment larvae had higher body weights and developed faster than the others. However, winter survival rates were significantly lower in the warm treatment compared with the cold treatment, and there were serious side effects of the higher temperature regime as development progressed. Warm conditions resulted in the lowest weights of pupae and adult females, together with shorter forewings in adult males, and earlier emergence of both sexes compared with the other treatments. We can only surmise as to whether these physiological effects of warm winters would translate into a primary cause of population decline, and whether they would apply to other species such as the Mountain Ringlet. Nevertheless, they reinforce a general point that warmer winters are unlikely to be good news for all British flora and fauna.

Mayflies are widely distributed in and around streams and rivers throughout the UK, but there is evidence of population declines in many places, and indications that warming of the waters inhabited by mayfly larvae may be a contributory factor. These insects have in the past been extraordinarily abundant, with clouds of adults emerging on warm spring days to the delight of anglers and naturalists alike.

Experiments suggest that Green Drakes are among the mayflies likely to suffer population declines as their freshwater habitats become warmer.

The larvae, by virtue of this luxuriance, are key components of the aquatic communities wherever they thrive. As gill-breathers, mayfly nymphs are very sensitive to dissolved oxygen concentrations, and this in turn is negatively affected by increasing temperature. Experiments by Verberk *et al.* (2016) were designed to determine the importance of this relationship for two common British mayflies, the Green Drake and the Blue-winged Olive *Serratella ignita*, and to compare the results with measurements made in the field where these species occur. In the laboratory, oxygen concentration and temperature were varied systematically and the effects on mayfly nymphs were recorded. Oxygen stress lowered lethal temperature limits by 5.5°C and 8.2°C for Green Drake and Blue-winged Olive larvae respectively. In the field, poor oxygenation was associated with greatly reduced site occupancy, especially under warm water conditions. Low oxygenation curtailed optimal stream temperatures for both species. Lab and field research therefore agreed, and pointed to serious consequences for mayflies if waters become warmer, exacerbated by eutrophication from farm run-off which further depletes dissolved oxygen in all too many streams and rivers.

Transplantation experiments into newly available climate space are among the most ambitious and imaginative responses to the problems of climate change. The successful establishment of Marbled White

Experimental translocation of Small Skippers demonstrated the validity of a climate envelope model for this butterfly.

and Small Skipper butterflies in habitats north of their previous range limit, as mentioned in Chapter 3, demonstrates the power of this approach as an important tool for future conservation. Very likely we will see more of it.

Vertebrate experiments

Species under threat from climate change have, reasonably enough, been the subject of experimentation much more often than those benefiting from it. Cold-water fishes such as Vendace and Powan are cases in point, as mentioned earlier and in Chapter 4. Some experiments to assess the influence of shortening winters on these species focused on embryonic development, from immediately post hatch to the start of feeding (Karjalainen *et al.*, 2015). These early stages in the lives of fishes are widely considered to be the most sensitive to environmental variation, and thus likely to provide clues about future impacts of climate change. The experiments involved incubating eggs under 'long', 'short' and 'shortest winter' conditions, the first corresponding to the current situation while the other two were derived from climate change projections. 'Shortening' of the winter period, corresponding to an increase in average temperature, reduced the development time of both species by up to twofold.

Survival, however, was unaffected by the trial conditions. Probably the most parsimonious interpretation of this experiment is that its basic hypothesis, that early developmental stages are the most sensitive to warmer water, was simply wrong. This result certainly does not detract from the value of the study, which can form the basis for future investigations into other aspects of Vendace and Powan physiology. It may well be that the computer modelling of Vendace ecology described earlier in this chapter identified the most sensitive issue for this fish, namely summer temperature and oxygen concentrations and their effects on adult survival.

Lichen experiments

Lichens make a substantial contribution to overall biodiversity at high altitudes, but the specialised, cold-tolerant species that live there may face a bleak future as climate change progresses. Dispersal is usually slow for this group of organisms and in most cases unlikely to keep up with the pace of predicted warming even if new sites come to exist within suitable climate envelopes.

For one arctic-alpine lichen, *Flavocetraria nivalis*, this problem has been addressed by a combination of modelling and transplant experiments undertaken by Brooker *et al.* (2016). Attempts were made to generate a habitat model based on the characteristics of its existing locations in the Cairngorm mountains. This model was then used to identify translocation sites, both fairly close to the lichen's current distribution and distant from it. Multiple transfers of *Flavocetraria nivalis* were carried out between 2010 and 2012, and their survival and growth were then monitored until 2015. The model, which worked to some extent when tested within the current range of *Flavocetraria nivalis*, performed poorly when used to explain survival and growth rates at the translocation sites.

Further work to improve the model is clearly needed, perhaps including more fine-scale measurements of microclimates where the lichens occur. Nevertheless, lichens did survive the transplantation process – and that discovery in itself was a valuable outcome of the project, which is evidently a work in progress. One interesting practical point was the realisation that lichens in unsuitable conditions can take years to expire. Long-term monitoring is essential to distinguish slow growth from lingering death.

New horizons and limits to predictive methods

Computer modelling is very much a science in progress, and developments continue apace. Pearce-Higgins *et al.* (2017) combined climate envelope assessments with past and recent distribution data for more than 3,000 plants and animals in Great Britain, and also generated risk assessments for each species to estimate the likely impacts of climate change. The outcome suggested that, under a moderate (low emissions) climate change scenario, perhaps 21 per cent of species are likely to suffer range reductions while rather more, maybe 38 per cent, will benefit from range extensions, although most of the predictions came with wide confidence limits. More upland species were at risk than lowland ones. These deductions are broadly in accord with, but more extensive and sophisticated than, most earlier studies.

Equipped with supercomputers and elegant laboratories, a scientific community unparalleled in history with respect to its size and expertise has set out to show us what the future of British wildlife will look like. How much confidence does this impressive enterprise carry with it? Certainly it is a better bet than prediction based on personal intuition, although the difference has not always been obvious. In 1978 one Leonard Koppett employed a system of his own devising to work the stock market. He correctly predicted market ups and downs for 18 of the next 19 years, presumably getting rich along the way. His method, when finally unveiled, was based on which of two football teams won the American Super Bowl championship. This kind of chancy, sustained success has been given a name: the hot-hand fallacy. It always runs out eventually.

The assessment of climate change impacts by Pearce-Higgins *et al.* (2017) found that upland habitats contained more species at risk than other habitats.

Nevertheless, computer-based fortune telling also has its limitations. In the 1960s, Edward Lorenz made an early attempt to forecast weather using a computer program. His groundbreaking discovery was not how to generate accurate predictions, but rather the converse. Extraordinarily trivial changes in the data fed into the program generated wildly different forecasts, a result that later became known as the 'butterfly effect'. The 'butterfly' problem is now widely recognised, and computer programs try to account for it by increasing the quantity of input data and by estimating error limits on whatever predictions emerge. Despite these valiant efforts it remains the case that unpredictable events make forecasts increasingly unreliable as time elapses.

In 1968, Paul and Anne Ehrlich published *The Population Bomb*, which anticipated a famine-driven apocalypse before the turn of the millennium on the basis that food supply would not keep up with the burgeoning human population. We are still here because the authors did not foresee, and could not have foreseen, the agricultural revolution that massively increased food production. Before howling with derision at what looks like a hopeless failure, it is wise to take a closer look at that exercise. The green revolution has entailed massive habitat destruction and degradation, untold numbers of species extinctions and the propagation of polluting pesticides on an industrial scale. Many people, including the Ehrlichs, suspect that the predictions were unrealised only because their timing was wrong. We now rely heavily on technological advances to support human numbers that continue to increase, with consequences that may yet emerge in a very unwelcome guise. In the early 1970s the Club of Rome, a group of prominent intellectuals, published *The Limits to Growth*. This computer-modelled assessment of possible futures included a wide range of variables, including population increase, that might unfold in the coming decades. A 30-year update concluded that, if anything, the situation with regard to global sustainability was worse than the various scenarios had anticipated.

The unfulfilled prophesy from the 1960s highlights the inevitably chaotic undercurrents in global events that can never be fully captured by the most elaborate computer model. Aside from this kind of unpredictability, there could be worse news to come. Studies of sunspots and solar output, again utilising computer models, are beginning to indicate a dramatic cooling that by 2030 could be inducing a mini ice age comparable with that experienced in the late Middle Ages. Will there be lobbies for an increase in fossil-fuel burning to protect us from biting cold in the not-too-distant future? The imponderables are huge, too big to be sure of anything beyond the next year or two, but despite that uncertainty we cannot ignore what is happening right now. Somehow we have to plan a response and try to protect our wildlife from climate change if that is at all possible, and hope we get it right. Progress along these lines is the subject matter of the final chapter.

Fish passes, as pictured here at Pitlochry, are vital auxiliaries to dams generating hydroelectric power.

impassable barriers for migratory fishes including salmon, sea trout, eels and lampreys. Salmon runs have become depleted or eventually disappeared altogether in the worst cases.

In more enlightened times, fish ladders have been constructed to allow migrants to pass the obstacle and gain access to their upstream spawning grounds, but these do not always restore the population to its former glory. Some systems work better than others. The fish pass at Pitlochry on the River Tummel functions effectively, and radio-tagged salmon that approached it all negotiated its multiple chambers successfully. Unfortunately, not all rivers have fared this well. Salmon entering the River Conon in northern Scotland have to endure a daunting set of obstacles including four fish passes if they are to reach their spawning redds. In this case, Gowans *et al.* (2003) noted that only 4 out of 54 tagged fish successfully moved through the complete system.

The Conon and associated rivers have maintained viable salmon populations, but it is not clear whether they would do so in the absence of the huge annual releases of juvenile fish from artificial hatcheries. Ancillary measures, including gates to prevent animals moving downstream from entering the turbines, are also essential. Inadequate guards can result in the deaths of up to two-thirds of the fish that bypass them and attempt to negotiate revolving turbine blades. Installing fish ladders is expensive, and even now many rivers remain inaccessible to migrants. The failure of recent attempts to reintroduce

Adult European Eels need safe passage downstream, avoiding hydroelectric turbines.

salmon to the River Thames has been attributed, at least in part, to the persistence of too many impassable weirs and dams.

After something of a hiatus, interest in creating more hydroelectric schemes in the UK has revived as part of the impetus for more renewable energy. It will be important to ensure that impacts on fishes are minimised in all new hydro schemes. River systems with substantial populations of salmon, sea trout, lampreys or eels, all of which are declining, should ideally be avoided altogether. But even in rivers of lesser importance for these species, where new developments go ahead the inclusion of efficient fish passes is still highly desirable. A widespread response to warming water has been for non-migratory fishes to move upstream to find favourable temperature conditions, a response that will be difficult or impossible if barriers without bypasses are imposed.

Hydroelectricity has long been present in the background, but it is the recent proliferation of wind turbines across the British countryside that has given the strongest signal of a shift away from fossil-fuel dependency. Marching across the uplands like serried ranks of war-of-the-worlds Martian robots, the glaring prominence of wind farms has made this response to climate change controversial. Landscapes are as inspiring a part of British heritage as the wildlife within them, and for many people their defacement by technology is a step too far. Carbon dioxide is a villain, but it does not spoil the view. An emerging response to this concern is to build wind farms offshore, some of which are now very extensive despite the extra cost involved. Wind is another energy source that the UK, with its face to the Atlantic, is well placed to exploit, and no doubt there will be more turbines in the years ahead.

Conservation in a warming world

Apart from aesthetic concerns, though, there are more specific problems to worry about with these powerful machines. Any kind of flying animal is at risk from being mangled by turbine rotor blades with tips moving at up to 300km per hour, and this happens on all too many occasions. Birds, especially wildfowl and raptors, have been killed, and sometimes in very large numbers, as they try to negotiate land-based wind farms. In the worst cases, such as around turbine arrays near Gibraltar that are bang in the middle of a major bird migration route, several thousand birds die every year, and bat mortality can be equally devastating. Even small, individual wind turbines cause fatal accidents, and the 20,000 or more units installed in the UK are estimated to account for the deaths of up to 5,000 birds and 3,000 bats annually. Offshore wind farms pose much lower risks to terrestrial-based birds and bats but transfer mortality threats to seabirds that forage near them. Around the British coastline, Gannets *Morus bassanus* look particularly vulnerable because they fly at heights within the range of turbine rotor blades, and extensive offshore wind farms are planned for areas close to major breeding sites. Predictions about the impact of these developments on Gannets are

ABOVE: Gannets are at risk from offshore wind farms.

BELOW: Badly placed wind turbines kill many birds and bats.

disturbing. Several thousand adult and juvenile birds may die, and, although the population is large, it is estimated that this extra mortality would be enough to precipitate a long-term decline.

Conservationists are not alone in their concerns about wind farms, and protests about planning applications for them are plentiful. Effigies of turbines have been burnt in Scotland, and the existence of a health condition, 'wind turbine syndrome', has been proposed as a basis for objecting to them. This is a 'psychosomatic' disorder, essentially a feeling of anxiety when near turbines. However, it is not recognised as a 'real' illness anywhere in the world. Feelings run high and have generated graphic, largely unrepeatable, comments in Australian public meetings, summarised by one disgruntled attendee: 'for wind energy the most appropriate language is profanity, vulgarity, and obscenity. The louder the better.'

It seems likely that, partly as a result of widespread protestations, new land-based wind farms will face tougher tests before being approved. But turbines will not go away, and wildlife conservation efforts have to work around that fact. Engagement with planners at early stages of new wind farm proposals is one way forward, already taken up by the RSPB, which examines hundreds of plans every year. The aim is to avoid major migration routes and important feeding, breeding and roosting areas of birds known or suspected to be at risk. This strategy can succeed, as it did when deterring construction of a wind farm on the Isle of Lewis that would have put Golden Eagles *Aquila chrysaetos*, among other species, in unacceptable danger.

Mitigating damage to bat populations is more difficult to prescribe, but not impossible. In Pennsylvania, increasing the cut-in speed of rotor blades (that is, the wind speed below which the machines do not operate) when bats were active reduced mortality by up to 93 per cent with a minimal loss of less than 1 per cent of power output. The dangers of offshore turbines could probably be reduced by raising the rotors above the usual foraging heights of birds such as Gannets. So far, at least, there is little to suggest that population declines of any bird or bat in the UK have been exacerbated by wind turbines, and in comparison with other causes of mortality they are usually small beer. Nevertheless, now is the time to be prudent and prevent an existing, sometimes unpleasant, situation developing into a more serious one. There may even be unplanned benefits from offshore wind farms, as they create regions of seabed relatively undisturbed by fishing activity, essentially constituting small marine reserves.

A third form of renewable energy generation for which the UK is also in pole position is tidal power. The British coastline is blessed with some of the highest tidal ranges in the world, 13m or more in the Bristol Channel, which is second only to Canada's Bay of Fundy. Moreover, unlike the wind, tides are completely reliable. Though lagging behind that of wind turbines, tide-harnessing technology has progressed considerably and could make increasingly large contributions to the UK green energy budget. Unfortunately realisation of this vast potential might, if the more extravagant schemes are adopted, prove hugely damaging to wildlife. Tidal barrages have been proposed for several estuaries, including that of the River Severn in the Bristol Channel. Submerged turbines produce energy during both the rise and fall of the tide but, as with hydroelectric generators in rivers, they pose threats to any fishes caught up in them. They also create major obstacles for migrating fishes, but the most damaging and inevitable result is the loss of mudflats used as feeding grounds by colossal numbers of waders and other water birds.

That is exactly what happened after completion of the Cardiff Bay barrage, which resulted in a large lake designed as an amenity and which does not generate power, at the turn of the millennium. The operation reduced both the amount of available habitat and the Shelduck *Tadorna tadorna* population in Cardiff Bay by 90 per cent, according to Ferns & Reed (2009). If, as has been suggested, a barrage were

Shelduck were almost exterminated in Cardiff Bay as a result of barrage construction.

constructed across the full width of the Bristol Channel, the intertidal habitat of some 85,000 birds would be irrevocably lost. The Severn estuary is recognised as a wetland area of international importance, designated as a Ramsar site (a Special Protection Area under the EC Directive on the Conservation of Wild Birds) and a Special Area of Conservation (SAC) under the EU Habitats Directive, and partly protected by Site of Special Scientific Interest (SSSI) status. Other estuaries caught in the sights of tidal power enthusiasts are also, very often, among the best for wildlife and enjoy similar levels of statutory protection. Hopefully all this legislative paraphernalia will give pause for thought. Harnessing tidal power does not have to be so destructive. Derivative plans for tidal lagoons, including one near Cardiff, raise the possibility of using tidal power on a practical but much less ecologically damaging scale.

Finally there is solar power, an energy source for which the UK is not as obviously suitable as it is for the three discussed above. Perhaps surprisingly, the prospects for capitalising on sunshine are much better than might be expected. British summers are warmer than they often seem, and insolation in the south of the country is comparable with that in Germany, which in 2015 generated about 7 per cent of its electricity from sunshine. Photoelectric units can be small enough to site on house roofs or, increasingly often, can be produced in arrays for solar farms in fields of a hectare or more in extent. Their contribution to UK energy demands increased rapidly after 2000, and 'reflective fields', looking from a distance like small lakes, increasingly adorn the British countryside. Like wind turbines, the landscape impact of these panels has not been universally welcomed, but solar power is less demanding of land area than any of the other methods for producing renewable energy.

Although solar farms seem benign to wildlife, there may be risks. Extensive arrays can fool birds, as well as humans, into perceiving them as water rather than as a solid substrate. This can make for a bumpy, perhaps lethal landing. The problem is even more acute for night-flying aquatic insects, especially water beetles, which regularly come to an ignominious end on greenhouse roofs. Devices that focus sunlight, such as those developed in the USA, are a different matter, and some have killed large numbers of birds by setting fire to them when they wander into areas of concentrated beams. No such machines are planned in the UK, and at present there is no reason to believe that conventional solar farms constitute a significant threat to any plants or animals. On the contrary, solar farms look like a good investment for the benefit of wildlife. Panels are raised above the ground, and more than 95 per cent of a solar-farm field remains accessible for plant growth, and thus for wildlife enhancements. Grazing by sheep is compatible with the technology, as is the creation of wildflower meadows around the solar panels.

Four sites in England have demonstrated how valuable sympathetic management of solar farms can be for wildlife conservation, as shown by Parker & McQueen (2013). All of them supported greater biodiversities than adjacent plots that were of the same habitat type as the solar farms prior to their construction. Those re-sown as wildflower meadows fared the best, with increases of grassland herbs, bumblebees and butterflies. Other wildlife was also seen on the farms during the survey, including rare birds, Brown Hares, signs of small mammals and a wide range of invertebrates. The challenge will be to make this kind of transition actually happen, ideally as conditions of planning permissions, in as many solar-farm sites as possible. Apart from benefiting wildlife, the dashes of summer colour reminiscent of a lost countryside might mute criticism by those concerned about a glass-fronted landscape.

There is no question that renewable energy must be harnessed if the overriding goal of reversing climate change is ever to happen. The challenge is to ensure that wildlife interests are centre-stage as the process unfolds. Green energy needs to be green in every respect.

Great diving beetles such as this female *Dytiscus marginalis* often mistake dry surfaces for water, and may fall victim to solar energy panels.

Solar farms are perhaps the greenest type of green energy production.

Sea-level rise

Another consequence of global warming against which the UK is fortifying itself is an inexorable rise in sea level. Increasingly high tides are intensifying erosion around soft coastlines, especially estuaries, and wearing away precious saltmarsh habitats. The oldest method for trying to contain threats from pounding waves is the construction of sea walls. Concrete barricades and earth embankments several metres wide, sometimes faced on the seaward side by stone or concrete blocks, had snaked their way round more than 2,000km of coastline in England and Wales by 2015. This amounts to 30 per cent of the entire coastal fringe in those two countries. In some areas, especially Essex, Suffolk, and around the Wash, Humber and Severn estuaries, hardly any unmodified shoreline remains, and pressure is on to extend this line of defence even further.

Sea walls are a mixed blessing for wildlife. They provide considerable areas of potentially valuable habitat, currently some 10,000 hectares altogether, safe from the agricultural pressures often manifest on land that adjoins them. When managed appropriately they support a wide range of wildlife and provide corridors between larger sites with good habitat, a feature that will be increasingly important for species attempting to change their distribution in response to climate change, all as discussed by Gardiner *et al.* (2015). However, hardening the coastline in this way has a downside, as mentioned in Chapter 7. Saltmarsh cannot retreat inland when sea walls are present, and the resulting habitat squeeze contributes to its loss. This is a serious matter for wading birds, including migrants, that rely on saltmarsh fauna as a food source. Other animals such as Natterjack Toads are also imperilled by the trend. Future extensions of sea walls are likely to prove contentious, and they should be resisted or modified where they threaten more saltmarsh, shingle or dune habitats.

A more recent response to threats from the sea has been the development of managed retreat, a strategy that sacrifices land, mostly agricultural, by removing sections of sea wall and permitting localised flooding on the landward side. From the wildlife perspective this is usually good news, because new saltmarsh can form as a result. Farmers may be less enthusiastic, and approvals for managed retreat involve lengthy negotiations that do not always succeed. In Sussex's Cuckmere valley, proposals to stop defending river embankments

Sea walls can provide good habitat for wildlife including Adders (left) but also add an extra squeeze to saltmarshes.

near the estuary dragged on for more than ten years as a result of local opposition. There are some risks from the wildlife perspective too, because areas where managed retreat looks like a realistic option sometimes include important freshwater habitats that would be badly damaged by saltwater inundation, as at Dungeness.

In the right place managed retreat can have positive results, as it has at Abbots Hall on the Essex coast. Breaching of an old sea wall in 2002 led to the creation of 80 hectares of saltmarsh, mudflats and coastal grassland. Birds including Brent Geese *Branta bernicla*,

Climate Change and British Wildlife

Managed retreat provides new habitat for Short-eared Owls.

Redshanks, Lapwings and Short-eared Owls *Asio flammeus* have subsequently thrived there, and the new creeks support populations of fish and marine invertebrates for which wading birds forage. This management method, when carefully targeted, represents a bonus for wildlife – and with any luck we will see much more of it in future.

Flood protection

One aspect of climate that has regularly made the news in the UK since the turn of the millennium is the rise in number and severity of torrential storms. Although climatologists are reluctant to ascribe individual incidents to climate change, these storms comply with one of the key climate model predictions, namely the likelihood of ever more extreme weather events. Floods caused by these huge downpours have been dramatic, drowning not only low-lying towns and villages but also large expanses of countryside.

Heightening of flood defence walls along river banks is an ongoing operation in the most vulnerable urban areas, but in the countryside a common call has been to dredge rivers and thus speed water on its way to the sea. The plea has often been emotional rather than rational. It sounds like a sensible idea, but dredging is expensive, may not work, and

Flooding has increased in wildlife habitats, such as the Somerset Levels.

can be problematic for wildlife. In the winter of 2013/14 the Somerset Levels experienced the largest flood ever known there, covering 65km^2 of low-lying farmland. Driven largely by public pressure rather than expert opinion, 8km of the rivers Parrett and Tone were subsequently dredged. Until the next flood we will not know whether this expensive operation will be effective. Low-lying areas like the Somerset Levels have always been susceptible to floods, and most of the wildlife in these places is there because the habitat suits them. Farmers and householders take a different view about the merits of occasional inundation – not unreasonably, since livelihoods and homesteads can be devastated by these dramatic weather episodes. Can the interests of wildlife and people be reconciled in situations such as this?

Unfortunately, the dredging of rivers certainly damages wildlife habitat. Removal of vegetation from within channels and along their banks reduces shade, increases water temperature and lowers oxygen concentrations. Invertebrate numbers and diversity are hit, with possible knock-on effects on predators including fishes and Otters *Lutra lutra*. The impact of dredging extends beyond the initial disturbance. Decreased soil stability along banks increases erosion and input of sediment, which can quickly obviate any depth increase caused by the dredging operation, and may also smother fish eggs. Dredging

River dredging as a response to flooding has downsides for wildlife.

destroys Water Vole *Arvicola amphibius* burrow systems and immediately afterwards increases the mortality rate of a rodent already of high conservation concern. The reduction in water levels in neighbouring fields associated with dredging can also cause declines in breeding birds such as Snipe and Redshank that require shallow flood pools.

From the wildlife point of view, therefore, dredging is far from an ideal solution to flood problems. Fortunately alternatives are available and, not before time, are being taken seriously by statutory agencies and landowners. One method is simply to set aside land that is allowed to flood when the need arises. The Ouse Washes in East Anglia are an old but excellent example of this strategy. Often submerged in winter and sometimes in summer, this fenland habitat provides a vast wetland much appreciated by wading birds while also minimising flood risk in the surrounding area. In Somerset, and hopefully elsewhere too, a holistic approach including not just the lowlands but the entire catchment area is being developed. The 'Hills to Levels' plan includes new tree planting, installation of leaky ponds and wooden dams and new hedge banks. Reducing the rate at which rainwater spills off the land is the main aim, allowing progressive drainage into streams and rivers and thus a more moderate flow between upland and estuary. No doubt dredging will continue to feature as a management practice, at least for a while, but in the long term this new thinking offers the best flood protection hope for humans and wildlife alike.

Recolonisation of the UK by Cranes started on strictly protected wildlife sites.

Options to assist climate change beneficiaries

Irrespective of any need to extend protected areas to compensate for changes in climate envelopes, there is a good case for also maintaining existing ones. Whooper Swans *Cygnus cygnus*, Goldeneyes *Bucephala clangula*, Little Egrets, Common Cranes *Grus grus*, Mediterranean Gulls *Larus melanocephalus* and Cetti's Warblers are wetland birds that have successfully colonised the UK since the 1960s. All of them began their new adventures in well-established protected sites, from which they subsequently spread far beyond. As more newcomers are expected, it will be important to keep these 'landing pads' available to help them establish.

Habitat management can improve the prospects of species that are already benefiting from climate change over and above what warming itself can achieve. Glanville Fritillary butterflies *Melitaea cinxia* frequent south-facing slopes and grasslands in southern England. Management designed to create a mosaic of short turf interspersed with taller vegetation will likely optimise conditions for this and other insects that reach their northern range margin within the UK.

The Silver-spotted Skipper is another warmth-loving butterfly extending its range in south-east England. In this case, increasing the survival prospects of individual populations promotes range expansion because colonies near range margins are at high risk of extinction from random events. Population viability can be improved through local management by enlarging habitat patch size and increasing its quality, and by improving connectivity between patches.

Managing the marine environment

Few options are available to moderate the consequences of a warming ocean. What is clear, though, is that the damaging effects of a warming climate can be exacerbated by human fishing activities. The North Sea sand eel fishery, in particular, is exaggerating the impact of climate change on British seabirds, especially Kittiwakes. The British sand eel fishery has largely been closed down, but an international response, especially ending the much larger Danish operation, is highly desirable now. Another useful measure would be the increase and expansion of marine nature reserves, designated as Marine Protected Areas (MPAs), where commercial fishing is limited or completely excluded. By 2016 about 23 per cent of British coastal waters were designated as MPAs. However, only a minority of these are specifically dedicated to wildlife protection, as Nature Conservation MPAs. More areas with this top-level protection, preferably linked as networks, could increase resilience to the impact of climate change on marine communities, for example by allowing increases in sand eel numbers.

Grand plans

Conservation organisations, both statutory and non-governmental, have begun wide-ranging responses to the threat of climate change. The RSPB has initiated activities across its reserves, which encompass examples of all the UK's important habitats, as described by Ausden (2014). These habitats vary considerably in terms of perceived risk. Coastal environments give by far the greatest cause for concern, while heaths, woods and upland grasslands look reasonably safe from climate change, at least in the immediate future. Intertidal zones, saline lagoons and fresh or brackish wetlands near the sea are particularly endangered.

These assessments have already been translated into actions. Intertidal zones have been modified by managed realignment, including alterations to coastal defences. Freshwater grazing marshes and reedbeds have been created at sites away from the coast,

Wallasea Island is one of the largest conservation projects in the UK, aiming to create new swathes of coastal wildlife habitat.

Conservation in a warming world

Creation of inland reedbeds, as here at Ouse Fen in Cambridgeshire, has, so far, more than compensated for damage to coastal habitats under threat from sea-level rise.

and some have already proved dramatically successful by, for example, attracting Bitterns and Cranes as at Lakenheath Fen on the Norfolk–Suffolk border. Artificial drains in peatlands and woodlands have been blocked to retain moisture, while grasslands have been managed to regulate water levels according to need in summer and winter.

The range of pre-emptive strikes by the UK's largest NGO involved in nature conservation hopefully paves the way for more of the same in the coming years. Some of the necessary measures take a long time to implement. Wallasea Island off the Essex coast has become a paradigm for coastal habitat restoration but, from a starting design in 2006, it took a full ten years before managed realignment and thus habitat improvement was actually under way. Inland reedbeds can take well over a decade to establish sufficiently to support the full range of birds that benefit from them, while drain blocks can produce habitat improvements within half that time. There are two messages here. One is that pioneering conservation management designed to ameliorate the impact of climate change can work, and work well. The second is that it takes a considerable time and, often, substantial financial resources. The clock is ticking and the wallet is stretched to the limit.

Future prospects

British wildlife is on a rising temperature escalator. Its pace has varied, getting off to a slow start in the 1980s, speeding up in the 1990s and slowing a little, but not stopping, in the early years of the 21st century. We have no way of knowing how near the top of the escalator we are now, nor what its future speed will be. Even if it stopped tomorrow, lags in the responses of our wildlife to climate changes that have already occurred mean it would be many years before stability is restored.

But of course the escalator will not stop tomorrow, or the day after. There is as yet no sign that global efforts to curb the burning of fossil fuels are biting heavily enough to slow the inexorable global temperature rise significantly. One trivial consequence of the relentlessly evolving situation is that this book can be no more than a snapshot of climate impacts on British wildlife. The escalator moves on up, and views from it will continue to shift erratically as it goes. Any temptation to dwell on the positive aspects of climate change needs careful thought. The 'gambler's ruin' is a paradigm playing out in the UK as the climate warms. Species doing well in the new environment can increase and spread without immediate restraint, though no matter how successful they become there is nothing to prevent them falling back later. By contrast, species suffering adverse effects of climate change may disappear forever without chance

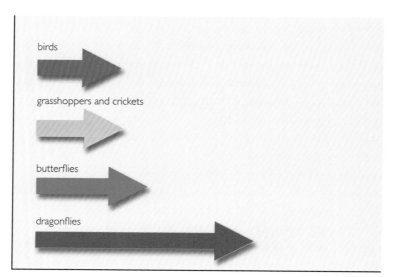

Relative trends in northerly range extensions in some mobile British animals. Lengths of arrows indicate relative average trends for each taxonomic group.

of resurgence if they fall to the 'absolute zero' state of extinction. The wheel of chance is unbalanced. Nor should we lose sight of the broader perspective. Climate warming is having much more profound effects on wildlife elsewhere in the world than anything under way in the British Isles. What chance for Polar Bears *Ursus maritimus* by the end of the 21st century?

As if it isn't difficult enough to preserve the UK's wildlife heritage, conservation organisations are faced with new uncertainties as the vagaries of climate change play out before us. Action must now take account not only of current crises but also of models that may, or may not, describe disasters yet to come. Understanding ecological problems sufficiently well to prescribe effective conservation management is not a game for the faint-hearted, even when the environment is more or less stable. There have been striking successes but also lessons of humility when nature scorns what seem like well-thought-out solutions. There will always be a joker in the pack, and planning for climate change will likely invigorate it. This comes at a time when protecting British wildlife is at an all-time low on the political agenda and financial support from government is declining faster than an endangered species.

The UK's forthcoming departure from the European Union adds another level of uncertainty, especially with respect to the future of Countryside Stewardship schemes. Fortunately, many of the methods necessary to manage for climate change are modifications or extensions of those long in place to deal with problems of the status quo. This gives confidence that at least in the short term, before climate alters beyond recognition, currently successful techniques should continue to work. Technologies developed to generate renewable energy, improve sea defences and reduce flood risk all have associated dangers for wildlife, but these look containable providing the will is there. Practical attempts to alleviate the expected impact of climate change are already under way and will need to intensify in future, but this will only happen on a sufficient scale if funding rises to the challenge. Pessimism is unbecoming, but it is difficult to refute Tim Sparks's take on where we have ended up:

> *What we may more clearly acknowledge is that our species and countryside have changed. Gone are almost all of our landscape elm trees, oilseed rape is a common crop, and many mixed farming areas are now almost totally arable. I grew up in Cambridgeshire, too late*

to see the last of the local Red Squirrels, and much has changed since then. In my boyhood Magpies, Buzzards and Red Kites would never be encountered while Song Thrushes and House Sparrows were much more common. Rarely now do I see Hares or Hedgehogs and I wouldn't even know where to see the former boxing in the month of March. I used to fish in a local stream for sticklebacks and freshwater shrimps. However, that stream now includes run-off from a dual carriageway and I can't be bothered to see what survives in its murky waters. It's all rather depressing. Do we have an idealised memory of the past, remembering only the Good Old Days, or have we witnessed some serious deterioration in wildlife and landscape in our lifetimes? From scientific research and monitoring we know the latter to be true. But there are also some success stories, which give us a modicum of hope for the world our children must inherit.

Central to extricating ourselves from this dilemma will be recognising the role that human numbers play in it. The UK is now home to far more people than are sustainable by a balanced and benign environment, and that is the fundamental reason why we haven't got one. Worries about food security have dominated agricultural policy since the Second World War, and the single most effective way to reduce greenhouse gas emissions would be to reduce the number of people dependent on the energy that gas production generates. This remains largely a taboo subject despite the best efforts of lobbyists such as Population Matters. Will there ever be an equivalent of the Paris Climate Change Summit, this time dedicated to humane and consensual methods for restraining population growth? Those itinerant wanderers from the age of ice and biting cold have, in the space of a few millennia, spread like a vast mycelium across a planet now aching under the stress of overwhelming human numbers. Awakening to this aftermath of unprecedented success is surely the key to a secure future for humans and wildlife alike. In the meantime we can only continue to press for more immediate measures to ensure that future generations will be able to take pleasure in the British countryside and its engaging wildlife. It is, after all, a magnificent legacy, as varied and spectacular as anywhere on the planet. We cannot, must not, squander it.

OPPOSITE PAGE:
Mist shrouds Glastonbury Tor during a spring sunrise on the Somerset Levels.

References

Akimova, A., Núñez-Riboni, I., Kempf, A. & Taylor, M. H. 2016. Spatially-resolved influence of temperature and salinity on stock and recruitment variability of commercially important fishes in the North Sea. *PLOS ONE* 11 (9): e0161917. doi: 10.1371/journal.pone.0161917.

Alberto, F., Bouffier, L., Louvet, J-M. *et al.* 2011. Adaptive responses for seed and leaf phenology in natural populations of sessile oak along an altitudinal gradient. *Journal of Evolutionary Biology* 24: 1442–1454.

Alsos, I. G., Alm, T., Normand, S. & Brochmann, C. 2009. Past and future range shifts and loss of diversity in dwarf willow (*Salix herbacea* L.) inferred from genetics, fossils and modeling. *Global Ecology and Biogeography* 18: 223–239.

Altermatt, F. 2010. Climatic warming increases voltinism in European butterflies and moths. *Proceedings of the Royal Society B* 277: 2009.1910. doi: 10.1098/rspb.2009.1910.

Amano, T., Freckleton, R. P., Queenborough, S. A. *et al.* 2014. Links between plant species' spatial and temporal responses to a warming climate. *Proceedings of the Royal Society B* 281: 20133017. doi: 10.1098/rspb.2013.3017.

Amano, T., Smithers, R. J., Sparks, T. H. & Sutherland, W. J. 2010. A 250-year index of first flowering dates and its response to temperature changes. *Proceedings of the Royal Society B* 277: 2451–2457.

Ausden, M. 2014. Climate change adaptation: putting principles into practice. *Environmental Management* 54: 685–698.

Austin, G. E. & Rehfisch, M. M. 2005. Shifting nonbreeding distributions of migratory fauna in relation to climatic change. *Global Change Biology* 11: 31–38.

Balfour-Browne, F. 1940. *British Water Beetles*. Ray Society.

Beale, C., Burfield, I. M., Innes, I. J. *et al.* 2006. Climate change may account for the decline in British ring ouzels *Turdus torquatus*. *Journal of Animal Ecology* 75: 826–835.

Beebee, T. J. C. 1995. Amphibian breeding and climate. *Nature* 374: 219–220.

Bell, J. R., Alderson, L., Izera, D. *et al.* 2015. Long-term phenological trends, species accumulation rates, aphid traits and climate: five decades of change in migrating aphids. *Journal of Animal Ecology* 84: 21–34.

Bell, T. 1849. *A History of British Reptiles*. J. Van Voorst, London.

Blockeel, T. L., Bosanquet, S. D. S., Hill, M. O. & Preston, C. D. (eds) 2014. *Atlas of British and Irish Bryophytes*, 2 volumes. Pisces, Newbury.

Bock, A. A., Sparks, T. H., Estrella, N. *et al.* 2014. Changes in first flowering dates and flowering duration of 232 plant species on the island of Guernsey. *Global Change Biology* 20: 3508–3519.

Britton, A. J., Beale, C. M., Towers, W. & Hewison, R. L. 2009. Biodiversity gains and losses: evidence for homogenisation of Scottish alpine vegetation. *Biological Conservation* 142: 1728–1739.

Brodie, J. F., Post, E. S. & Doak, D. F. (eds) 2012. *Wildlife Conservation in a Changing Climate*. University of Chicago Press, Chicago.

Brooker, R.W., Brewer, M., Britton, A.J. *et al.* 2016. Feasibility study: translocation of species for the establishment or protection of populations in northerly and/or montane environments. *Scottish Natural Heritage Commissioned Report No. 913*.

Brooks, S., Parr A. & Mill, P. 2007. Dragonflies as climate change indicators. *British Wildlife* 19: 85–93.

Burns, F., Eaton, M. A., Barlow, K. E. *et al.* 2016. Agricultural management and climatic change are the major drivers of biodiversity change in the UK. *PLOS ONE* 11(3): e0151595. doi: 10.1371/journal.pone.0151595.

Burthe, S., Butler, A., Searle, K. R. et al. 2011. Demographic consequences of increased winter births in a large aseasonally breeding mammal (*Bos taurus*) in response to climate change. *Journal of Animal Ecology* 80: 1134–1144.

Burton, J. F. & Sparks, T. H. 2002. Flying earlier in the year: the phenological responses of butterflies and moths to climate change. *British Wildlife* 13: 305–311.

Carroll, E. A., Sparks, T. H., Collinson, N. & Beebee, T. J. C. 2009. Influence of temperature on the spatial distribution of first spawning dates of the common frog (*Rana temporaria*) in the UK. *Global Change Biology* 15: 467–473.

Carroll, M. J., Dennis, P., Pearce-Higgins, J. W. & Thomas, C. D. 2011. Maintaining northern peatland ecosystems in a changing climate: effects of soil moisture, drainage and drain blocking on craneflies. *Global Change Biology* 17: 2991–3001.

Cathrine, C. 2014. Grass snakes in Scotland. *The Glasgow Naturalist* 26: 36–40.

Chapman, D. S. 2013. Greater phenological sensitivity to temperature on higher Scottish mountains: new insights from remote sensing. *Global Change Biology* 19: 3463–3471.

Chapman, D. S., Bell, S., Helfer, S. & Roy, D. B. 2015. Unbiased inference of plant flowering phenology from biological recording data. *Biological Journal of the Linnean Society* 115: 543–554.

Clare, F. C., Halder, J. B., Daniel, O. et al. 2016. Climate forcing of an emerging pathogenic fungus across a montane multi-host community. *Philosophical Transactions of the Royal Society B* 371: 20150454. doi: 10.1098/rstb.2015.0454.

Clegg, J. 1959. *The Freshwater Life of the British Isles*, 2nd edition. Frederick Warne, London.

Club of Rome. 1972. *The Limits to Growth*. Earth Island Ltd.

Comte, L. & Grenouillet, G. 2013. Do stream fish track climate change? Assessing distribution shifts in recent decades. *Ecography* 36: 1236–1246.

Coudun, C. & Gégout, J-C. 2007. Quantitative prediction of the distribution and abundance of *Vaccinium myrtillus* with climatic and edaphic factors. *Journal of Vegetation Science* 18: 517–552.

Crick, H. Q. P. & Sparks, T. H. 1999. Climate change related to egg-laying trends. *Nature* 399: 434–424.

D'Andrea, L., Broennimann, O., Kozlowsk, G. et al. 2009. Climate change, anthropogenic disturbance and the northward range expansion of *Lactuca serriola* (Asteraceae). *Journal of Biogeography* 36: 1573–1587.

Davey, C. M., Chamberlain, D. E., Newson, S. E., Noble, D. G. & Johnston, A. 2012. Rise of the generalists: evidence for climate driven homogenization in avian communities. *Global Ecology and Biogeography* 21: 568–578.

Doxford, S. W. & Freckleton, R. P. 2012. Changes in the large-scale distribution of plants: extinction, colonisation and the effects of climate. *Journal of Ecology* 100: 519–529.

Dulvy, N. K., Rogers, S. I., Jennings, S. et al. 2008. Climate change and deepening of the North Sea fish assemblage: a biotic indicator of warming seas. *Journal of Applied Ecology* 45: 1029–1039.

Ehrlich, P. 1968. *The Population Bomb*. Ballantine Books, New York.

Eiseley, L. 1969. *The Unexpected Universe*. Harcourt, Brace & World, New York.

Elliott, J. A. & Bell, V. A. 2011. Predicting the potential long-term influence of climate change on vendace (*Coregonus albula*) habitat in Bassenthwaite Lake, U.K. *Freshwater Biology* 56: 395–405.

Ellis, C. 2015. Implications of climate change for UK bryophytes and lichens. *Biodiversity Climate Change Impacts Report Card Technical Paper 8*. Royal Botanic Garden, Edinburgh.

Ellis, C., Coppins, B. J., Dawson, T. P. & Seaward, M. R. D. 2007. Response of British lichens to climate change scenarios: trends and uncertainties in the projected impact for contrasting biogeographic groups. *Biological Conservation* 140: 217–235.

Emerson, R. W. 1876. *Letters and Social Aims*. The Riverside Press, Cambridge, Massachusetts.

Evans, P. G. H. & Bjørge, A. 2013. Impacts of climate change on marine mammals. *MCCIP Science Review* 2013: 134–148.

Ewald, J., Wheatley, C. J., Aebischer, N. J. et al. 2014. Cereal invertebrates, extreme events and long-term trends in climate. *Natural England Commissioned Report* NECR135.

Ferns, P. N. & Reed, J. P. 2009. Effects of the Cardiff Bay tidal barrage on the abundance, ecology and behaviour of shelducks *Tadorna tadorna*. *Aquatic Conservation* 19: 466–473.

Fitter, A. H. & Fitter, R. S. R. 2002. Rapid changes in flowering time in British plants. *Science* 296: 1689–1691.

Fox, R., Oliver, T. H., Harrower, C. et al. 2014. Long-term changes to the frequency of occurrence of British moths are consistent with opposing and synergistic effects of climate and land-use changes. *Journal of Animal Ecology* 51: 949–957.

Frederiksen, M., Anker-Nilssen, T., Beaugrand, G. & Wanless, S. 2013. Climate, copepods and seabirds in the boreal Northeast Atlantic: current state and future outlook. *Global Change Biology* 19: 364–372.

Frederiksen, M., Elston, D. A., Edwards, M., Mann, A. D. & Wanless, S. 2011. Mechanisms of long-term decline in size of lesser sandeels in the North Sea explored using a growth and phenology model. *Marine Ecology Progress Series* 432: 137–147. doi: 10.3354/meps09177.

Frederiksen, M., Wanless, S., Harris, M. P., Rothery, P. & Wilson, L. J. 2004. The role of industrial fisheries and oceanographic change in the decline of North Sea black-legged kittiwakes. *Journal of Applied Ecology* 41: 1129–1139.

Gardiner, T. 2009. Macropterism of Roesel's bushcricket *Metrioptera roeselii* in relation to climate change and landscape structure in eastern England. *Journal of Orthoptera Research* 18: 95–102.

Gardiner, T., Pilcher, R. & Wade, M. 2015. *Sea Wall Biodiversity Handbook*. RPS, UK.

Gillingham, P. K., Bradbury, R. B., Roy, D. B. et al. 2015. The effectiveness of protected areas in the conservation of species with changing geographical ranges. *Biological Journal of the Linnean Society* 115 (3): 707–717.

Gowans, A. R. D., Armstrong, J. D., Priede, I. G. & Mckelvey, S. 2003. Movements of Atlantic salmon migrating upstream through a fish-pass complex in Scotland. *Ecology of Freshwater Fish* 12: 177–189.

Green, R. E., Collingham, Y. C., Willis, S. G. et al. 2008. Performance of climate envelope models in retrodicting recent changes in bird population size from observed climatic change. *Biology Letters* 4: 599–602.

Griffiths, R. A., McCrea, R. S. & Sewell, D. 2010. Dynamics of a declining amphibian metapopulation: survival, dispersal and the impact of climate. *Biological Conservation* 143: 485–491.

Groom, Q. J. 2013. Some poleward movements of native vascular plants is occurring but the fingerprint of climate change is not evident. *PeerJ* 1: e77. doi: 10.7717/peerj.77.

Hassall, C. & Thompson, D. J. 2010. According for recorder effort in the detection of range shifts from historical data. *Methods in Ecology and Evolution* 1: 343–350.

Hassall, C., Thompson, D. J., French, G. C. & Harvey, I. F. 2007. Historical changes in the phenology of British Odonata are related to climate. *Global Change Biology* 13: 933–941.

Hayhow, D. B., Ausden, M. A., Bradbury, R. B. et al. 2017. *The State of the UK's Birds 2017*. RSPB, BTO, WWT, DAERA, JNCC, NE and NRW, Sandy, Bedfordshire.

Hickling, R., Roy, D. B., Hill, J. K., Fox, R. & Thomas, C. D. 2006. The distributions of a wide range of taxonomic groups are expanding polewards. *Global Change Biology* 12: 450–455.

Hill, M. O. & Preston, C. D. 2015. Disappearance of boreal plants in southern Britain: habitat loss or climate change? *Biological Journal of the Linnean Society* 115: 598–610.

Hill, M. O. & Preston, C. D. 2014. Changes in distribution and abundance, 1960–2013. In Blockeel, T. L., Bosanquet, S. D. S., Hill, M. O. & Preston, C. D. (eds), *Atlas of British and Irish Bryophytes, Volume 1*: 34–49. Pisces, Newbury.

Hinks, A. E., Cole, E. F., Daniels, K. J. et al. 2015. Scale-dependent phenological synchrony between songbirds and their caterpillar food source. *The American Naturalist* 186: 84–97.

Hinsley, S. A., Bellamy, P. E., Hill, R. A. & Ferns, P. N. 2016. Recent shift in climate relationship enables prediction of the timing of bird breeding. *PLOS ONE* 11 (5): e0155241. doi: 10.1371/journal.pone.0155241.

Hudson, I. L. & Kealey, M. R. 2010. *Phenological Research: Methods for Environmental and Climate Change Analysis*. Springer, New York.

Hudson, W. H. 1903. *Hampshire Days*. Longman, London.

Hulme, P. E. 2011. Contrasting impacts of climate-driven flowering phenology on changes in alien and native plant species distributions. *New Phytologist* 189: 272–281.

Jackson, D. W. T. & Cooper, J. A. G. 2011. Coastal dune fields in Ireland: rapid regional response to climatic change. *Journal of Coastal Research* SI 64: 293–297.

Jaeschke, A., Bittner, T., Jentsch, A. et al. 2012. Biotic interactions in the face of climate change: a comparison of three modelling approaches. *PLOS ONE* 7(12): e51472. doi: 10.1371/journal.pone.0051472.

Jaeschke, A., Bittner, T., Reineking, B. & Beierkuhnlein, C. 2013. Can they keep up with climate change? Integrating specific dispersal abilities of protected Odonata in species distribution modelling. *Insect Conservation and Diversity* 6: 93–103.

Jeffries, R. 1879. *Wild Life in a Southern County*. Smith, Elder & Co., London.

Jones, G., Barlow, K., Ransome, R. & Gilmour, L. 2015. *Greater Horseshoe bats and their insect prey: the impact and importance of climate change and agri-environment schemes*. University of Bristol, Bristol.

Jones, M. C., Dye, S. R., Fernandes, J. A. et al. 2013. Predicting the impact of climate change on threatened species in UK waters. *PLOS ONE* 8(1): e54216. doi: 10.1371/journal.pone.0054216.

Karjalainen, J., Keskinen, T., Pulkkanen, M. & Marjomäki, T. J. 2015. Climate change alters the egg development dynamics in cold-water adapted coregonids. *Environmental Biology of Fishes* 98: 979–991.

Kerr, J. T., Pindar, A., Galpern, P. et al. 2015. Climate change impacts on bumblebees converge across continents. *Science* 349: 177–180.

Kerr, R. A. 2009. What happened to global warming? Scientists Say Just Wait a Bit. *Science* 326: 28–29.

Kery, M., Dorazio, R. M., Soldaat, L., van Strien, A., Zuiderwijk, A. & Royle, J. A. 2009. Trend estimation in populations with imperfect detection. *Journal of Applied Ecology* 46: 1163–1172.

Kington, J. A. 2010. *Climate and Weather*. New Naturalist 115. Collins, London.

Kirby, K. J., Smart, S. M., Black, H. I. J. et al. 2005. Long term ecological change in British woodland (1971–2001). *English Nature Research Reports No. 653*.

Lundy, M., Montgomery, I. & Russ, J. 2010. Climate change-linked range expansion of Nathusius' pipistrelle bat, *Pipistrellus nathusii* (Keyserling & Blasius, 1839). *Journal of Biogeography* 37: 2232–2242.

Macdonald, D. W. & Newman, C. 2002. Population dynamics of badgers (*Meles meles*) in Oxfordshire, UK: numbers, density and cohort life histories, and a possible role of climate change in population growth. *Journal of Zoology* 256: 121–138.

Marren, P. 2012. *Mushrooms*. British Wildlife Publishing, Gillingham, Dorset.

Martay, B., Brewer, M. J., Elston, D. A. et al. 2017. Impacts of climate change on national biodiversity population trends. *Ecography* 40: 1139–1151. doi: 10.1111/ecog.02411.

Mason, S. C., Palmer, G., Fox, R. et al. 2015. Geographical range margins of many taxonomic groups continue to shift polewards. *Biological Journal of the Linnean Society* 115 (3): 586–597.

Massimino, D., Johnston, A. & Pearce-Higgins, J. W. 2015. The geographical range of British birds expands during 15 years of warming. *Bird Study* 62: 523–534.

May, R. 1986. Biological Diversity: how many species are there? *Nature* 324: 514–515.

McGrath, A. L. & Lorenzen, K. 2010. Management history and climate as key factors driving natterjack toad population trends in Britain. *Animal Conservation* 13: 483–494.

Mieszkowska, N., Hawkins, S. J., Burrows, M. T. & Kendall, M. A. 2007. Long-term changes in the geographic distribution and population structures of *Osilinus lineatus* (Gastropoda: Trochidae) in Britain and Ireland. *Journal of the Marine Biological Association of the UK* 87: 537–545.

Milazzo, M., Cattano, C., Alonzo, S. H. et al. 2016. Ocean acidification affects fish spawning but not paternity at CO_2 seeps. *Proceedings of the Royal Society B* 283: 20161021. doi: 10.1098/rspb.2016.1021.

Moss, B. 2014. Freshwaters, climate change and UK nature conservation. *Freshwater Reviews* 7: 25–75.

Moyes, K., Nussey, D. H., Clements M. N. et al. 2011. Advancing breeding phenology in response to environmental change in a wild red deer population. *Global Change Biology* 17: 2455–2469.

Newnham, R. M., Sparks, T. H., Skjoth, C. A., Head, K., Adams-Groom, B. & Smith, M. 2013. Pollen season and climate: is the timing of birch pollen release in the UK approaching its limit? *International Journal of Biometeorology* 57: 391–400.

Ohlberger, J., Thackeray, S. J., Winfield, I. J., Maberly, S. C. & Vøllestad, L. A. 2014. When phenology matters: age-size truncation alters population response to trophic mismatch. *Proceedings of the Royal Society B* 281: 20140938. doi: 10.1098/rspb.2014.0938.

O'Neill, B. F., Bond, K., Tyne, A. et al. 2012 Climatic change is advancing the phenology of moth species in Ireland. *Entomologia Experimentalis et Applicata* 143: 74–88. doi: 10.1111/j.1570-7458.2012.01234.x.

Pakanen, V-M., Orell, M., Vatka, E., Rytkönen, S. & Broggi, J. 2016. Different ultimate factors define timing of breeding in two related species. *PLOS ONE* 11(9): e0162643. doi: 10.1371/journal.pone.0162643.

Palmer, G., Platts, P. J., Brereton, T., Chapman, J. W. et al. 2017. Climate change, climatic variation and extreme biological responses. *Philosophical Transactions of the Royal Society B* 372: 20160144. doi: 10.1098/rstb.2016.0144.

Parker, G. E. & McQueen, C. 2013. Can solar farms deliver significant benefits to biodiversity? Preliminary study July–August 2013. Unpublished report, Wychwood Biodiversity.

Parr, A. J. 2010. Monitoring of Odonata in Britain and possible insights into climate change. *Biorisk* 5: 127–139.

Pateman R. M., Thomas, C. D., Hayward, S. A. L. & Hill, J. K. 2016. Macro- and microclimatic interactions can drive variation in species' habitat associations. *Global Change Biology* 22: 556–566.

Pearce-Higgins, J. W. 2017. Birds and climate change. *British Birds* 110: 388–404.

Pearce-Higgins, J. W., Beale, C. M., Oliver, T. H. et al. 2017. A national-scale assessment of climate change impacts on species: assessing the balance of risks and opportunities for multiple taxa. *Biological Conservation* 213: 124–134.

Pearce-Higgins, J. W. & Green, R. E. 2014. *Birds and Climate Change: Impacts and Conservation Responses*. Cambridge University Press, Cambridge.

Reading, C. J. 2007. Linking global warming to amphibian declines through its effects on female body condition and survivorship. *Oecologia* 151: 125–131.

Reed, T. E., Jenouvrier, S. & Visser, M. E. 2013. Phenological mismatch strongly affects individual fitness but not population demography in a woodland passerine. *Journal of Animal Ecology* 82: 131–144.

Robbirt, K. M., Roberts, D. L., Hutchings M. J. & Davy, A. J. 2014. Potential disruption of pollination in a sexually deceptive orchid by climate change. *Current Biology* 24: 2485–2849.

Root, T. L., Hall, K. R., Herzog, M. P. & Howell, C. A. 2015. *Biodiversity in a Changing Climate: linking science and management in conservation*. University of California Press, Oakland, CA.

Ross, L. 2015. Climate change impacts on the vegetation of Ben Lawers. *Scottish Natural Heritage Commissioned Report No. 879*.

Roy, D. B., Oliver, T. H., Botham, M. S. et al. 2015. Similarities in butterfly emergence dates among populations suggest local adaptation to climate. *Global Change Biology* 21: 3313–3322.

Ruiz-Navarro, A., Gillingham, P. K. & Britton, J. R. 2016. Predicting shifts in the climate space of freshwater fishes in Great Britain due to climate change. *Biological Conservation* 203: 33–42.

Schneider, S. H. & Root, T. (eds) 2002. *Wildlife Responses to Climate Change*. Island Press, Washington, DC.

Sedlacek, J., Wheeler, J. A., Cortés, A. J. et al. 2015. The response of the alpine dwarf shrub *Salix herbacea* to altered snowmelt timing: lessons from a multisite transplant experiment. *PLOS ONE* 10 (4): e0122395. doi: 10.1371/journal.pone.0122395.

Slingsby, D., Hopkins, J., Carter, S., Dalrymple, S. & Slingsby, A. 2010. Change and Stability: Monitoring the Keen of Hamar 1978–2006. *Scottish Natural Heritage report*.

Smale, D. A., Wernberg, T., Yunnie, A. L. E. & Vance, T. 2015. The rise of *Laminaria ochroleuca* in the western English Channel (UK) and comparisons with its competitor and assemblage dominant *Laminaria hyperborea*.

Marine Ecology 36: 1033–1044.

Smith, M. 1951. *The British Amphibians & Reptiles*. Collins, London.

Sparks, T. H. & Carey, P. D. 1995. The responses of species to climate over two centuries: an analysis of the Marsham phenological record, 1736–1947. *Journal of Ecology* 83: 321–329.

Sparks, T. H., Collinson, N., Crick, H. et al. 2006. Natural heritage trends of Scotland: phenological indicators of climate change. *Scottish Natural Heritage Commissioned Report No. 167* (ROAME No. F01NB01).

Sparks, T. & Smithers, R. 2009. The blurring of the seasons: a focus on autumn phenology. *British Wildlife* 20: 247–249.

Stewart, J. A., Bantock, T. M., Beckmann, B. C. et al. 2015. The role of ecological interactions in determining species ranges and range changes. *Biological Journal of the Linnean Society* 115: 647–663.

Stewart, J. A. & Kirby, P. 2010. Hemiptera. In Maclean, N. (ed.), *Silent Summer: the State of Wildlife in Britain and Ireland*. Cambridge University Press, Cambridge.

Stuhldreher, G., Hermann, G. & Fartmann, T. 2014. Cold-adapted species in a warming world: an explorative study on the impact of high winter temperatures on a continental butterfly. *Entomologia Experimentalis et Applicata* 151: 270–279.

Taylor, A., Bruine de Bruin, W. & Dessai, S. 2014. Climate change beliefs and perceptions of weather-related changes in the United Kingdom. *Risk Analysis* 34: 1995–2004. doi: 10.1111/risa.12234.

Telfer, M. G., Preston, C. D. & Rothery, P. 2002. A general method for measuring relative change in range size from biological atlas data. *Biological Conservation* 107: 99–109.

Thackeray, S. J., Henrys, P. A., Hemming, D. et al. 2016. Phenological sensitivity to climate across taxa and trophic levels. *Nature* 535: 241–245.

Thackeray, S. J., Jones, I. D. & Maberly, S. C. 2008. Long-term change in the phenology of spring phytoplankton: species-specific responses to nutrient enrichment and climatic change. *Journal of Ecology* 96: 523–535.

Thackeray, S. J., Sparks, T. H., Frederiksen, M. et al. 2010. Trophic level asynchrony in rates of phenological change for marine, freshwater and terrestrial environments. *Global Change Biology* 16: 3304–3313.

Thomas, C. D. & Lennon, J. J. 1999. Birds extend their ranges northwards. *Nature* 399: 213.

van der Meer, S., Jacquemyn, H., Carey, P. D. & Jongejans, E. 2016. Recent range expansion of a terrestrial orchid corresponds with climate-driven variation in its population dynamics. *Oecologia* 181: 435–448.

Verberk, W. C. E. P., Durance, I., Vaughan, I. P. & Ormerod, S. J. 2016. Field and laboratory studies reveal interacting effects of stream oxygenation and warming on aquatic ectotherms. *Global Change Biology* 22: 1769–1778.

Webster, E., Cameron, R. W. F. & Culham, A. 2017. *Gardening in a Changing Climate*. Royal Horticultural Society, London.

Weinert, M., Mathis, M., Kroncke, I. et al. 2016. Modelling climate change effects on benthos: distributional shifts in the North Sea from 2001 to 2099. *Estuarine, Coastal and Shelf Science* 175: 157–168.

Wethay, D. S., Woodin, S. A., Hilbish, T. J. et al. 2011. Response of intertidal populations to climate: effects of extreme events versus long term change. *Journal of Experimental Marine Biology and Ecology* 400: 132–144.

White, G. 1789. *The Natural History & Antiques of Selborne*. White, Cochrane & Co., London.

Wiles, E. 2012. Farmers' perception of climate change and climate solutions. *Briefing Note 1*, Global Sustainability Institute, Anglia Ruskin University.

Williams, J. L., Jacquemin, H., Ochoki, B. M., Brys, R. & Miller, T. E. X. 2015. Life history evolution under climate change and its influence on the population dynamics of a long-lived plant. *Journal of Ecology* 103: 798–808.

Woodland Trust. *Nature's Calendar*. https://naturescalendar.woodlandtrust.org.uk (accessed January 2018).

Zerefos, C. S., Gerogiannis, V. T., Balis, D., Zerefos, S. C. & Kazantzidis, A. 2007. Atmospheric effects of volcanic eruptions as seen by famous artists and depicted in their paintings. *Atmospheric Chemistry and Physics* 7: 4027–4042.

Abbreviations

BBS:	Breeding Bird Survey
BSBI:	Botanical Society of Britain and Ireland
bTB	bovine tuberculosis
BTO:	British Trust for Ornithology
CBC:	Common Birds Census
CEH:	Centre for Ecology and Hydrology
CET:	Central England Temperature
DNA:	Deoxyribonucleic Acid
ECN:	Environmental Change Network
EVI:	Enhanced Vegetation Index (for satellite-based analyses)
FAD:	First Arrival Date (of migratory birds)
FBA:	Freshwater Biological Association
HadCET:	Hadley Centre Central England Temperature (data set)
HadCM3:	Hadley Centre Coupled Model version 3
IPCC:	Intergovernmental Panel on Climate Change
MAD:	Median Arrival Date (of migratory birds)
MLD:	Median Laying Date (of egg-laying by birds)
MONARCH:	Modelling Natural Resource Responses to Climate Change
MPA:	Marine Protected Area
MTCI:	MERIS Terrestrial Chlorophyll Index (for satellite-based analyses)
NAO:	North Atlantic Oscillation (a correlate of winter weather conditions)
NBN:	National Biodiversity Network
NDVI:	Normalised Difference Vegetation Index (for satellite-based analyses)
NERC:	Natural Environment Research Council
NGO:	Non-government Organisation
pH:	measure of acidity
RHS:	Royal Horticultural Society
RIS:	Rothamsted Insect Survey
RSPB:	Royal Society for the Protection of Birds
SAC:	Special Area of Conservation
SOS	Start of growing season
SSSI:	Site of Special Scientific Interest
UKBMS:	UK Butterfly Monitoring Scheme
UKPN:	UK Phenology Network
WWF:	Worldwide Fund for Nature

Species names

Common English and scientific names are listed in alphabetical order of the English names, where these exist. Some species, for example of lichens and water beetles, only have scientific names. These are therefore listed under the group name (for example, 'Lichens').

Acorn barnacles
   – *Balanus crenatus*
   – *Semibalanus balanoides*
Adder *Vipera berus*
Adonis Blue *Polyommatus bellargus*
Alder *Alnus glutinosa*
Algae
   – *Asterionella formosa*
   – *Bifurcaria bifurcata*
   – *Cylotella*
   – *Delesseria sanguinea*
   – *Laminaria hyperborea*
   – *Laminaria ochroleuca*
   – *Membranoptera alata*
   – *Neodenticula seminae*
   – *Phycodrys rubens*
   – *Pleurococcus*
   – *Volvox*
Alpine Bearberry *Arctostaphylos alpina*
Alpine Catchfly *Lychnis alpina*
Alpine Gentian *Gentiana nivalis*
Alpine Haircap Moss *Polytrichastrum alpinum*
Alpine Newt *Ichthyosaura alpestris*
Amberjacks *Seriola* species
Anchovy *Engraulis encrasicolus*
Anemone Prawn *Periclimenes sagittifer*
Angelshark *Squatina squatina*
Angler Monkfish *Lophius piscatorius*
Ant
   – *Lasius neglectus*
Aphid
   – *Utamphorophora humboldti*

Arctic Charr *Salvelinus alpinus*
Arctic Sandwort *Arenaria norvegica*
Arctic Tern *Sterna paradisaea*
Ash *Fraxinus excelsior*
Asian Hornet *Vespa velutina*
Atlantic Cod *Gadus morhua*
Atlantic Horse Mackerel *Trachurus trachurus*
Atlantic Mackerel *Scomber scombrus*
Atlantic Salmon *Salmo salar*
Azure Hawker *Aeshna caerulea*

Bacteria
   – *Leptothrix ochracea*
   – *Mycobacterium bovis*
   – *Nostoc*
Badger *Meles meles*
Balding Pincushion *Ulota calvescens*
Banded Darter *Sympetrum pedomontanum*
Banded Demoiselle *Calopteryx splendens*
Baneberry *Actaea spicata*
Barn Owl *Tyto alba*
Barnacle Goose *Branta leucopsis*
Bar-tailed Godwit *Limosa lapponica*
Beaver *Castor fiber*
Beech *Fagus sylvatica*
Beetles
   – *Agabus bipustulatus*
   – *Cercyon sternalis*
   – *Dytiscus circumflexus*
   – *Dytiscus marginalis*
   – *Enochrus melanocephalus*
   – *Gyrinus urinator*
   – *Hydroporus scalesianus*
   – *Hydrovatus cuspidatus*
   – *Hygrotus nigrolineatus*
   – *Ilybius chalconatus*
   – *Laccobius simulatrix*
   – *Limnebius crinifer*
   – *Nebrioporus canaliculatus*

— *Noterus clavicornis*
— *Ochthebius alpinus*
— *Rhantus grapii*
— *Rhantus suturalis*
Beewolf *Philanthus triangulum*
Bell Heather *Erica cinerea*
Beluga Whale *Delphinapterus leucas*
Bib (Pouting) *Trisopterus luscus*
Bilberry *Vaccinium myrtillus*
Bilberry Bumblebee *Bombus monticola*
Bittern *Botaurus stellaris*
Blackbird *Turdus merula*
Blackcap *Sylvia atricapilla*
Black-faced Blenny *Tripterygion delaisi*
Black-footed Limpet *Patella depressa*
Black Grouse *Tetrao tetrix*
Black-tailed Skimmer *Orthetrum cancellatum*
Blackthorn *Prunus spinosa*
Blainville's Beaked Whale *Mesoplodon densirostris*
Blue Heath *Phyllodoce caerulea*
Blue-rayed Limpet *Patella pellucida*
Blue Tit *Cyanistes caeruleus*
Blue-winged Olive *Serratella ignita*
Bluebell *Hyacinthoides non-scripta*
Bog Bush-cricket *Metrioptera brachyptera*
Bowhead Whale *Balaena mysticetus*
Box *Buxus sempervirens*
Box Bug *Gonocerus acuteangulatus*
Bracken *Pteridium aqulinum*
Bramble *Rubus fruticosus*
Brent goose *Branta bernicla*
Brimstone *Gonepteryx rhamni*
Brittle stars
— *Acrocnida brachiata*
— *Amphiura filiformis*
— *Ophiothrix fragilis*
Broad-bodied Chaser *Libellula depressa*
Broom Fork-moss *Dicranum scoparium*
Brown Argus *Aricia agestis*
Brown Hare *Lepus europaeus*
Brown Hawker *Aeshna grandis*
Brown Rat *Rattus norvegicus*
Brown Trout *Salmo trutta*
Bryony Ladybird *Henosepilachna argus*
Bryozoans
— *Celloporella hyalina*
— *Electra pilosa*
Buddleia *Buddleja davidii*

Buffish Mining Bee *Andrena nigroaenea*
Bugs
— *Arenocoris falleni*
— *Athysanus argentarius*
— *Chorosoma schillingi*
— *Naucoris maculatus*
— *Odontoscelis lineola*
— *Sigara iactans*
— *Sigara longipalpis*
Burbot *Lota lota*
Buzzard *Buteo buteo*

California Mussel *Mytilus californianus*
Carnation Sedge *Carex panacea*
Carrion Crow *Corvus corone*
Cattle Egret *Bubulcus ibis*
Cetti's Warbler *Cettia cetti*
Chaffinch *Fringilla coelebs*
Chicory *Cichorium intybus*
Chiffchaff *Phylloscopus collybita*
Chillingham Cattle *Bos taurus*
Chinese Mitten Crab *Eriocheir sinensis*
Chub *Squalius cephalus*
Cinnamon Bug *Corizus hyoscyami*
Cloudberry *Rubus chamaemorus*
Clouded Yellow *Colias croceus*
Coal Tit *Periparus ater*
Cock's-foot *Dactylis glomerata*
Colt's-foot *Tussilago farfara*
Comma *Nymphalis c-album*
Common Blue *Polyommatus icarus*
Common Bream *Abramis brama*
Common Carp *Cyprinus carpio*
Common Crane *Grus grus*
Common Dog-violet *Viola riviniana*
Common Frog *Rana temporaria*
Common Limpet *Patella vulgata*
Common Marbled Carpet *Dysstroma truncata*
Common Poppy *Papaver rhoeas*
Common Quaker *Orthosia cerasi*
Common Reed *Phragmites australis*
Common Rush *Juncus effusus*
Common Sandpiper *Actitis hypoleucos*
Common Seal (Harbour) *Phoca vitulina*
Common Sedge *Carex nigra*
Common Skate *Dipturus batis*
Common Tern *Sterna hirundo*
Common Toad *Bufo bufo*

Common Yellow-sedge *Carex demissa*
Common Wasp *Vespula vulgaris*
Copepods
 – *Calanus finmarchicus*
 – *Calanus helgolandicus*
Corncockle *Agrostemma githago*
Cornflower *Centaurea cyanus*
Couch's Sea Bream *Pagrus pagrus*
Cowberry *Vaccinium vitis-idaea*
Creeping Willow *Salix repens*
Crucian Carp *Carassius carassius*
Cuckoo *Cuculus canorus*
Cuckooflower *Cardamine pratensis*
Cuckoo Ray *Leucoraja naevus*
Curled Hook-moss *Palustriella commutata*
Curlew *Numenius arquata*
Curlew Sandpiper *Calidris ferruginea*
Cuvier's Beaked Whale *Ziphius cavirostris*

Dabberlocks *Alaria esculenta*
Daddy-longlegs Spider *Pholcus phalangioides*
Dandelion *Taraxacum officinale*
Dark-barred Twin-spot Carpet *Xanthorhoe ferrugata*
Dark Blood Bee *Sphecodes niger*
Dartford Warbler *Sylvia undata*
Daubenton's Bat *Myotis daubentonii*
Deergrass *Trichophorum germanicum*
Dingy Skipper *Erynnis tages*
Dipper *Cinclus cinclus*
Dog Rose *Rosa canina*
Dorset Heath *Erica ciliaris*
Dover Sole *Solea solea*
Downy Birch *Betula pubescens*
Dunlin *Calidris alpina*
Dunnock *Prunella modularis*
Dusky Fork-moss *Dicranum fuscescens*
Dusty Miller *Cerastium tomentosum*
Dwarf Sperm Whale *Kogia sima*
Dwarf Willow *Salix herbacea*

Early Dog-violet *Viola reichenbachiana*
Early Spider-orchid *Ophrys sphegodes*
Edible Frog *Pelophylax esculentus*
Elder *Sambucus nigra*
Emperor Dragonfly *Anax imperator*
European Eel *Anguilla anguilla*
European Hornet *Vespa crabro*

European Plaice *Pleuronectes platessa*
European Sprat *Sprattus sprattus*

Fairy Ring Champignon *Marasmius oreades*
Fairy Shrimp *Chirocephalus diphanus*
Fallow Deer *Dama dama*
False Brome *Brachypodium sylvaticum*
False Morel *Gyomitra esculenta*
Fen Orchid *Liparis loeselii*
Field Maple *Acer campestris*
Fieldfare *Turdus pilaris*
Fingered Cowlwort *Colura calyptrifolia*
Fir Clubmoss *Huperzia selago*
Flat Topshell *Gibbula umbilicalis*
Flies
 – *Epistrophe melanostoma*
 – *Platypalpus aliterolamellatus*
 – *Sphegina sibirica*
 – *Tephritis cometa*
 – *Tephritis divisa*
Fly Agaric *Amanita muscaria*
Fragrant Orchid *Gymnadenia conopsea*
Fraser's Dolphin *Lagenodelphis hosei*
Frogspawn Alga *Batrachospermum torfusum*
Fungi
 – *Armillaria solidipes*
 – *Batrachochytrium dendrobatidis*
 – *Hymenoscyphus fraxineus*
 – *Morchella esculenta*
 – *Perenniporia ochroleuca*
 – *Phellinus wahlbergii*
 – *Puccinia graminis*
 – *Pulcherricium caeruleum*
 – *Saprolegnia*

Gannet *Morus bassanus*
Garden Bumblebee *Bombus hortorum*
Garden Carpet *Xanthorhoe fluctuata*
Garden Snail *Cornu aspersum*
Garden Tiger *Arctia caja*
Garlic Mustard *Alliaria petiolata*
Gatekeeper *Pyronia tithorus*
German Wasp *Vespula germanica*
Gilthead Bream *Sparus aurata*
Girdled Snail *Hygromia cinctella*
Glanville Fritillary *Melitaea cinxia*
Glittering Wood-moss *Hylocomium splendens*
Golden Eagle *Aquila chrysaetos*

Golden Plover *Pluvialis apricaria*
Goldeneye *Bucephala clangula*
Goosander *Mergus merganser*
Grass Snake *Natrix helvetica*
Grayling (butterfly) *Hipparchia semele*
Grayling (fish) *Thymallus thymallus*
Great Crested Newt *Triturus cristatus*
Great Lettuce *Lactuca virosa*
Great Silver Water Beetle *Hydrophilus piceus*
Great Tit *Parus major*
Great White Egret *Ardea alba*
Greater Horseshoe Bat *Rhinolophus ferrumequinum*
Green Drake Mayfly *Ephemera danica*
Green-flowered Helleborine *Epipactis phyllanthes*
Green Hawker *Aeshna viridis*
Green Shieldbug *Palomena prasina*
Green-veined White *Pieris napi*
Greenfinch *Chloris chloris*
Grey Bush-cricket *Platycleis albopunctata*
Grey Heron *Ardea cinerea*
Grey Plover *Pluvialis squatarola*
Grey Seal *Halichoerus grypus*
Grey Squirrel *Sciurus carolinensis*
Grey Topshell *Gibbula cineraria*
Grey Triggerfish *Balistes capriscus*
Gudgeon *Gobio gobio*
Guillemot *Uria aalge*

Hairy Dragonfly *Brachytron pratense*
Halibut *Hippoglossus hippoglossus*
Harbour Porpoise *Phocoena phocoena*
Harebell *Campanula rotundifolia*
Harlequin Ladybird *Harmonia axyridis*
Hawthorn *Crataegus monogyna*
Hazel *Corylus avellana*
Heath Bedstraw *Galium saxatile*
Heath-grass *Danthonia decumbens*
Heath Rush *Juncus squarrosus*
Hebrew Character *Orthosia gothica*
Hedgehog *Erinaceus europaeus*
Hen Harrier *Circus cyaneus*
Hermit crab *Pagurus pridaeux*
Herring *Clupea harengus*
Hoary Whitlowgrass *Draba incana*
Holly *Ilex aquifolium*
Holly Blue *Celastrina argiolus*
Honey Bee *Apis mellifera*
Honeysuckle *Lonicera perclymenum*

Hornet Mimic Hoverfly *Volucella zonaria*
Horse-chestnut *Aesculus hippocastanum*
Horse-chestnut Leaf Miner *Cameraria ohridella*
House Martin *Delichon urbicum*

Irish Daltonia *Daltonia splachnoides*
Ivy *Hedera helix*
Ivy-leaved Toadflax *Cymbalaria muralis*

Jersey Tiger *Eupalagia quadripunctaria*
Jumping plant louse
 – *Cacopsylla moscovita*

Keeled Skimmer *Orthetrum coerulescens*
Kestrel *Falco tinnunculus*
Kittiwake *Rissa tridactyla*
Knot *Calidris canutus*
Knotted-thread Hydroid *Obelia geniculata*

Lady Orchid *Orchis purpurea*
Lapwing *Vanellus vanellus*
Larch *Larix decidua*
Large Blue *Phengaris arion*
Large Heath *Coenonympha tullia*
Large Red Damselfly *Pyrrhosoma nymphula*
Large Tortoiseshell *Nymphalis polychloros*
Large White *Pieris brassicae*
Leatherback Turtle *Dermochelys coriacea*
Lesser Celandine *Ranunculus ficaria*
Lesser Emperor *Anax parthenope*
Lesser Gooseberry Sea Squirt *Distomus variolosus*
Lesser Marsh Grasshopper *Chorthippus albomarginatus*
Lesser Sand Eel *Ammodytes tobianus*
Lesser Silver Water Beetle *Hydrochara caraboides*
Lesser Weever Fish *Echiichthys vipera*
Lichens
 – *Flavocetraria nivalis*
 – *Flavoparmelia soredians*
 – *Lecanora populicola*
 – *Ochrolechia frigida*
 – *Pyrenula macrospora*
 – *Vulpicida pinastri*
Lilac *Syringa vulgaris*
Ling *Molva molva*
Ling Heather *Calluna vulgaris*
Little Bittern *Ixobrychus minutus*
Little Egret *Egretta garzetta*

Little Ringed Plover *Charadrius dubius*
Liverwort
 – *Mylia anomala*
Lizard Orchid *Himantoglossum hircinum*
Long-stalked Yellow-sedge *Carex lepidocarpa*
Long-winged Conehead *Conocephalus discolor*
Lugworm *Arenicola marina*
Lynx *Lynx lynx*

Madame's Pixie-cup Lichen *Cladonia coccifera*
Mallard *Anas platyrhynchos*
Man Orchid *Orchis anthropophora*
Marbled White *Melanargia galathea*
Marsh Forklet-moss *Dichodontium palustre*
Marsh Frog *Pelophylax ridibundus*
Mat-grass *Nardus stricta*
Meadow Foxtail *Alopecurus pratensis*
Meadow Pipit *Anthus pratensis*
Median Wasp *Dolichovespula media*
Mediterranean Gull *Larus melanocephalus*
Megrim *Lepidorhombus whiffiagonis*
Midges
 – *Culicoides imicola*
 – *Culicoides obsoletus*
 – *Culicoides pulicaris*
Midwife Toad *Alytes obstetricans*
Migrant Hawker *Aeshna mixta*
Minute Pouncewort *Cololejeunea minutissima*
Mistletoe *Viscum album*
Mole Cricket *Gryllotalpa gryllotalpa*
Monarch *Danaus plexippus*
Montagu's Stellate Barnacle *Chthamalus montagui*
Mountain Hare *Lepus timidus*
Mountain Ringlet *Erebia epiphron*
Mouse Moth *Amphipyra tragopoginis*

Narwhal *Monodon monoceros*
Nathusius' Pipistrelle *Pipistrellus nathusii*
Natterjack Toad *Bufo (Epidalea) calamita*
New Zealand Flatworm *Arthurdendyus triangulatus*
Nightingale *Luscinia megarhynchos*
Noble False Widow Spider *Steatoda nobilis*
Northern Brown Argus *Aricia artaxerxes*
Northern Damselfly *Coenagrion hastulatum*
Northern Dung Beetle *Agoliinus lapponum*
Northern Emerald *Somatochlora arctica*
Norway Pout *Trisopterus esmarkii*

Ocellated Wrasse *Symphodus ocellatus*
Orange Tip *Anthocharis cardamines*
Otter *Lutra lutra*
Oxeye Daisy *Leucanthemum vulgare*
Oystercatcher *Haematopus ostralegus*

Pacific Oyster *Crassostrea gigas*
Painted Lady *Vanessa cardui*
Palmate Newt *Lissotriton helveticus*
Peacock *Nymphalis io*
Pedunculate Oak *Quercus robur*
Perch *Perca fluviatilis*
Pheasant *Phasianus colchicus*
Pied Flycatcher *Ficedula hypoleuca*
Pike *Esox lucius*
Pill Sedge *Carex pilulifera*
Pine Martin *Martes martes*
Plane Tree Bug *Arocatus longiceps*
Pochard *Aythya ferina*
Pohlia Moss *Pohlia nutans*
Polar Bear *Ursus maritimus*
Polecat *Mustela putorius*
Poli's Stellate Barnacle *Chthamalus stellatus*
Pool Frog *Pelophylax lessonae*
Powan *Coregonus lavaretus*
Powdery mildews
 – *Microsphaera alphitoides*
 – *Microsphaera azaleae*
Prickly Goldenfleece *Urospermum picroides*
Prickly Lettuce *Lactuca serriola*
Ptarmigan *Lagopus muta*
Puffin *Fratercula arctica*
Purple Heron *Ardea purpurea*
Purple Moor-grass *Molinia caerulea*

Quail *Coturnix coturnix*

Rabbit *Oryctolagus cuniculus*
Racomitrium Moss *Racomitrium heterostichum*
Razorbill *Alca torda*
Red Admiral *Vanessa atalanta*
Red Blenny *Parablennius ruber*
Red Deer *Cervus elaphus*
Red Fox *Vulpes vulpes*
Red Helleborine *Cephalanthera rubra*
Red Kite *Milvus milvus*
Red Mullet *Mullus surmuletus*
Red Squirrel *Sciurus vulgaris*

Red-stemmed Feather-moss *Pleurozium schreberi*
Red-tailed Bumblebee *Bombus lapidarius*
Red-veined Darter *Sympetrum fonscolombii*
Redshank *Tringa totanus*
Redstart *Phoenicurus phoenicurus*
Redwing *Turdus iliacus*
Reed Warbler *Acrocephalus scirpaceus*
Reeves Muntjac Deer *Muntiacus reevesi*
Rigid Applemoss *Philonotis rigida*
Ring Ouzel *Turdus torquatus*
Ringed Plover *Charadrius hiaticula*
Ringlet *Aphantopus hyperantus*
Roach *Rutilus rutilus*
Robin *Erithacus rubecula*
Rock Sedge *Carex saxatilis*
Roesel's Bush-cricket *Metrioptera roeselii*
Rough Hawk's-beard *Crepis biennis*
Round-mouthed Snail *Pomatias elegans*
Rowan *Sorbus aucuparia*
Rudd *Scardinius erythrophthalmus*
Ruddy Darter *Sympetrum sanguineum*
Ruffe *Gymnocephalus cernua*
Rusty Swan-neck Moss *Campylopus flexuosus*

Sanderling *Calidris alba*
Sand Lizard *Lacerta agilis*
Sand Martin *Riparia riparia*
Sand Ray *Leucoraja circularis*
Sardine *Sardina pilchardus*
Sawfly Orchid *Ophrys tenthredinifera*
Saxon Wasp *Dolichovespula saxonica*
Scaldfish *Arnoglossus laterna*
Scarlet Darter *Crocothemis erythraea*
Scotch Argus *Erebia aethiops*
Scots Pine *Pinus sylvestris*
Screech Beetle *Hygrobia hermanni*
Sea-buckthorn *Hippophae rhamnoides*
Sea Daffodil *Pancratium maritimum*
Sea Potato *Echinocardium cordatum*
Selfheal *Prunella vulgaris*
September Thorn *Ennomos erosaria*
Sessile Oak *Quercus petraea*
Seven-spot Ladybird *Coccinella 7-punctata*
Shag *Phalacrocorax aristotelis*
Sheep Tick *Ixodes ricinus*
Shelduck *Tadorna tadorna*
Shetland Mouse-ear *Cerastium nigrescens*
Short-beaked Common Dolphin *Delphinus delphis*

Short-eared Owl *Asio flammeus*
Short-winged Conehead *Conocephalus dorsalis*
Signal Crayfish *Pacifastacus leniusculus*
Silky Forklet-moss *Dicranella heteromalla*
Silver Birch *Betula pendula*
Silver Bream *Blicca bjoerkna*
Silver-spotted Skipper *Hesperia comma*
Silver-studded Blue *Plebejus argus*
Siskin *Spinus spinus*
Slender Groundhopper *Tetrix subulata*
Slow-worm *Anguis fragilis*
Slugs
   – *Limacus flavus*
   – *Limacus maculatus*
Small-leaved Lime *Tilia cordata*
Small Red-eyed Damselfly *Erythromma viridulum*
Small Skipper *Thymelicus sylvestris*
Small Tortoiseshell *Nymphalis urticae*
Small White *Pieris rapae*
Smooth Lady's-mantle *Alchemilla glabra*
Smooth Newt *Lissotriton vulgaris*
Smooth Snake *Coronella austriaca*
Snake Pipefish *Entelurus aequoreus*
Snipe *Gallinago gallinago*
Snow-bed Willow (Dwarf Willow) *Salix herbacea*
Snow Bunting *Plectrophenax nivalis*
Snowdon Lily *Gagea serotina*
Snowdrop *Galanthus nivalis*
Song Thrush *Turdus philomelos*
Southern Damselfly *Coenagrion mercuriale*
Southern Emerald Damselfly *Lestes barbarus*
Southern Hawker *Aeshna cyanea*
Southern Marsh-orchid *Dactylorhiza praetermissa*
Southern Oak Bush-cricket *Meconema meridionale*
Spanish Bluebell *Hyacinthoides hispanica*
Sparrowhawk *Accipiter nisus*
Speckled Bush-cricket *Leptophyes punctatissima*
Speckled Wood *Pararge aegeria*
Spiny Dogfish (Spurdog) *Squalus acanthias*
Spiders
   – *Segestria florentina*
Spotted Flycatcher *Muscicapa striata*
Spotted Medick *Medicago arabica*
Spring Crocus *Crocus vernus*
Star Sedge *Carex echinata*
Starry Saxifrage *Saxifraga stellaris*
Stiff Sedge *Carex bigelowii*

Species names

Stinging Nettle *Urtica dioica*
Stone-curlew *Burhinus oedicnemus*
Stripe-winged Grasshopper *Stenobothrus lineatus*
Striped Dolphin *Stenella coeruleoalba*
Swallow *Hirundo rustica*
Swallowtail *Papilio machaon*
Sweet Vernal-grass *Anthoxanthum odoratum*
Swift *Apus apus*
Sycamore *Acer pseudoplantus*

Tadpole Shrimp *Triops cancriformis*
Ten-spined Stickleback *Pungitius pungitius*
Tench *Tinca tinca*
Three-spined Stickleback *Gasterosteus aculeatus*
Timothy *Phleum pratense*
Tongue Orchid *Serapias lingua*
Toothed Topshell *Osilinus lineatus*
Trailing Azalea *Loiseleuria procumbens*
Tree Bumblebee *Bombus hypnorum*
Treecreeper *Certhia familiaris*
Tufted Hair-grass *Deschampsia cespitosa*
Turtle Dove *Streptopelia turtur*
Twiggy Spear-moss *Warnstorfia sarmentosa*

Upland Summer Mayfly *Ameletus inopinatus*

Vendace *Coregonus albula*
Violet Carpenter Bee *Xylocopa violacea*
Viviparous Lizard *Zootoca vivipara*

Wall Butterfly *Lassiommata megera*
Wall Lizard *Podarcis muralis*
Wart Barnacle *Verruca stroemia*
Wasp Spider *Argiope bruennichi*
Water-soldier *Stratiotes aloides*
Water Vole *Arvicola amphibius*
Wavy Hair-grass *Deschampsia flexuosa*

Wheatear *Oenanthe oenanthe*
White-beaked Dolphin *Lagenorhynchus albirostris*
White-clawed Crayfish *Austropotamobius pallipes*
White Dead-nettle *Lamium album*
White Egg Fungus *Amanita ovoidea*
White-faced Darter *Leucorrhinia dubia*
White Frostwort *Gymnomitrion concinnatum*
White Henbane *Hyoscyamus albus*
White-legged Damselfly *Platycnemis pennipes*
Whitethroat *Sylvia communis*
Whiteworm Lichen *Thamnolia vermicularis*
Whooper Swan *Cygnus cygnus*
Wigeon *Anas penelope*
Wild Boar *Sus scrofa*
Willow Emerald Damselfly *Chalcolestes viridis*
Willow Warbler *Phylloscopus trochilus*
Winter Moth *Operophtera brumata*
Witch's Hair Lichen *Alectoria nigricans*
Wolf *Canis lupus*
Wolffish *Anarhichas lupus*
Wood Anemone *Anemone nemorosa*
Wood Avens *Geum urbanum*
Wood Cranesbill *Geranium sylvaticum*
Wood-sorrel *Oxalis acetosella*
Woodland Grasshopper *Omocestus rufipes*
Woodland Ringlet *Erebia medusa*
Wren *Troglodytes troglodytes*
Wryneck *Jynx torquilla*

Yellow-legged Mining Bee *Andrena flavipes*
Yellow Sole (Solenette) *Buglossidum luteum*
Yellowing Curtain Crust *Stereum subtomentosum*
Yew *Taxus baccata*
Yorkshire-fog *Holcus lanatus*

Zander *Sander lucioperca*
Zebra Mussel *Dreissena polymorpha*

Credits

All photographs are © the author, except for those listed below. Bloomsbury Publishing would like to thank those listed for providing photographs and for permission to reproduce copyright material within this book. While every effort has been made to trace and acknowledge all copyright holders, we would like to apologise for any errors or omissions, and invite readers to inform us so that corrections can be made in any future editions.

IMAGES
Inside back cover flap © Robert Rogers; 1 © Pauline Lewis/Getty Images; 2–3 © Moorefam/Getty Images; 10 © Mark Phillips/Alamy Stock Photo; 12 © Heritage Images/Getty Images; 15 © Julian Baker (JB Illustrations), based on data sourced from the Met Office; 16 top © Bob Gibbons, bottom © Grzegorz Petrykowski/Shutterstock.com; 17 © Paul Glendell/Alamy Stock Photo; 18 © Bob Gibbons; 19, 20, 23 © Laurie Campbell; 26 © Julian Baker (JB Illustrations); 28 top © James Lowen, bottom © Roger Key; 29 © Julian Baker (JB Illustrations); 32 © James Lowen; 33 © Laurie Campbell, inset © Martin Fowler/Shutterstock.com; 34 © Nigel Voaden; 36 © Roger Key; 38 © James Lowen; 41, 42, 45, 47, 49 © Laurie Campbell; 50 © Stewart Smith Photography/Shutterstock.com; 51 © Laurie Campbell; 52 © Lars Eklundh; 54 © Laurie Campbell; 55 left © Laurie Campbell, right © Bob Gibbons; 56 left © James Lowen, right © lidialongobardi77/Shutterstock.com; 57, 58 © Bob Gibbons; 59 left © Andrew Cleave/Nature Photographers, right © Laurie Campbell; 60 © Laurie Campbell; 61 © Mike Pennington, inset © Bob Gibbons; 63, 65 © James Lowen; 66 © Bob Gibbons; 71, 72 © Laurie Campbell; 74, 76 © James Lowen; 77 © Laurie Campbell; 78 © Bob Gibbons; 80 © Gary K Smith/Alamy Stock Photo; 82 © Marcel Jancovic/Shutterstock.com; 83 © Laurie Campbell; 85 © Rothamsted Research; 88 © Laurie Campbell; 89 © Nigel Voaden; 90 top © Nigel Voaden, bottom © James Lowen; 95 top © Nigel Voaden, bottom © David Goddard; 96 © Bob Gibbons; 97 © Julian Baker (JB Illustrations), adapted from Parr (2010); 98 © James Lowen; 99 © Laurie Campbell; 101 © Rothamsted Research; 103 © IanRedding/Shutterstock.com; 104 © Bob Gibbons; 105 © James Lowen; 106 © Steven Falk; 107 © James Lowen; 108 © Nigel Voaden; 109 © Phillip Buckham-Bonnett; 110 © Ant Cooper/Shutterstock.com; 111 © James Lowen; 112 © Jonty Denton; 113 © James Lowen; 115 top © Nigel Cattlin/Alamy Stock Photo, bottom © Roger Key; 118 © Bob Gibbons; 121 © Laurie Campbell; 123 © Erni/Shutterstock.com; 125 © Philip Birtwistle/Shutterstock.com, inset © Erni/Shutterstock.com; 128 © Laurie Campbell; 129 © Michael J. Hammett/Nature Photographers; 133 © Roger Wilmshurst/Getty Images; 137 © Laurie Campbell; 139 © Chris Gleed-Owen; 140 © IanRedding/Shutterstock.com; 141 © Michael Conrad/Shutterstock.com; 142 © Laurie Campbell; 145 top © Martin Fowler/Shutterstock.com, bottom © Julian Baker (JB Illustrations), adapted from Hinsley et al. (2016); 147 © James Lowen; 148, 149 © Nigel Voaden; 150, 151 © Laurie Campbell; 152 © Bob Gibbons; 153 top © Martin Fowler/Shutterstock.com, bottom © Nigel Voaden; 154 top © James Lowen, bottom © Nigel Voaden; 155, 156 © Laurie Campbell; 157 top © Laurie Campbell, bottom © Scott S. Brown/Shutterstock.com; 158 © Laurie Campbell; 160 © Bob Gibbons; 161 © James Lowen; 162 © Bob Gibbons; 165 top © Bob Gibbons, bottom © Laurie Campbell; 166 © Laurie Campbell; 167 left © fedsax/Shutterstock.com, right © Laurie Campbell; 168 © Paul Sterry/Nature

Credits

Photographers; 169, 170 © Laurie Campbell; 172 © sauletas/Shutterstock.com; 173 © Laurie Campbell; 175 © Steven Ellingson/Shutterstock.com; 176 © David Mark/Alamy Stock Photo; 177 © Joel Sartore/Getty Images; 179 © Jean-Yves Sgro; 180 © Laurie Campbell; 183 © camera lucida wildlife/Alamy Stock Photo; 185 © Laurie Campbell; 187 top © Erni/Shutterstock.com, bottom © Paul Cecil; 189, 190 © Laurie Campbell; 191 © Bob Gibbons; 192 © Laurie Campbell; 193 top © Laurie Campbell, bottom © Bob Gibbons; 194 © Nigel Voaden; 195 © Bob Gibbons; 196 © George Peterken; 197 © Laurie Campbell; 199 top © Gucio_55/Shutterstock.com, bottom © Nigel Voaden; 200 © Julian Baker (JB Illustrations), adapted from Pakanen et al. (2016); 201 © David Tipling/Getty Images; 202 top © Laurie Campbell, bottom © James Lowen; 204 © Laurie Campbell; 205 top © Sefton Library Service, bottom © Jason Smalley Photography/Alamy Stock Photo; 206 © FLPA/Hugh Clark/Getty Images; 207 © Bob Gibbons; 209 left © James Lowen, right © Nick Owens; 211 © Caroline Fitton; 212 © Laurie Campbell, inset © Nigel Voaden; 214, 216 © Laurie Campbell; 217 © Nigel Voaden; 218 © James Lowen; 219 © ECOSTOCK/Shutterstock.com; 220, 223 © Laurie Campbell; 224 © Mark Thomas; 225 © Paul Sterry/Nature Photographers; 227 top © Vladimir Wrangel/Shutterstock.com, bottom © S.C.Wild/Alamy Stock Photo; 228 © GoranStimac/Getty Images; 229 © Vladimir Wrangel/Shutterstock.com; 231 © Nigel Voaden; 232–234 © Laurie Campbell; 235 © Mark Harding/robertharding/Getty Images; 236 © wildestanimal/Getty Images; 237 © The Photolibrary Wales/Alamy Stock Photo; 238 © Linda George/Shutterstock.com; 241 © Jeff Morgan 13/Alamy Stock Photo; 244 © Chronicle/Alamy Stock Photo; 246 © Evening Standard/Getty Images; 247 top © Ian Rutherford/Alamy Stock Photo, bottom © Paul J Martin/Shutterstock.com; 248 © Topical Press Agency/Getty Images; 251 © Lilly Trott/Shutterstock.com; 252 top © Nick Hawkes/Shutterstock.com, bottom © EMJAY SMITH/Shutterstock.com; 254 © Allkindza/iStock.com; 255 © Jia Shao/Tim Sparks; 256 © Alice Thomas; 257 © BTO; 258 top © Jill Pelto, bottom © Charlotte Sullivan; 259 left © Photo 12/Getty Images, right © Granger Historical Picture Archive/Alamy Stock Photo; 260 © IanBeechImages; 262 top © Frederic Legrand – COMEO/Shutterstock.com, bottom © Joel Pett/2016 New York Times News Service; 263 © Peter Macdiarmid/Getty Images; 265 © pxl.store/Shutterstock.com; 266 © goodluz/Shutterstock.com; 269 © Crown copyright, Met Office, with permission granted under the terms of the Open Government Licence; 270 © Laurie Campbell; 271 © John Cancalosi/Getty Images; 273 © Paul Sterry/Nature Photographers; 275 © Israel Hervas Bengochea/Shutterstock.com; 278 © blickwinkel/Alamy Stock Photo; 280 © Martin Fowler/Shutterstock.com; 281 © Peter Guess/Shutterstock.com; 283 © Ivan Krawchuk/Shutterstock.com; 284 © Vladimir Wrangel/Shutterstock.com; 289 © Chris Ellis; 291 © Vladimir Wrangel/Shutterstock.com; 293, 294, 295 © Laurie Campbell; 301, 302 © James Lowen; 304 © Lorne Gill/SNH; 306 © Erni/Shutterstock.com; 309 top © Julian Baker (JB Illustrations), bottom © Heather Stuckey/RSPB Images; 310 © Julian Baker (JB Illustrations), adapted from Burns et al. (2016); 311 © Neil Mitchell/Shutterstock.com; 312 © Laurie Campbell; 313 © Jim Barton; 314 © jack perks/Alamy Stock Photo; 315 top © Laurie Campbell, bottom © J. Marijs/Shutterstock.com; 317 © Nigel Voaden; 319 top © Susan & Allan Parker/Alamy Stock Photo, bottom © zsolt_uveges/Shutterstock.com; 321 top © Tim Gardiner, bottom © Laurie Campbell; 322 © Nigel Voaden; 324 © Patrizio Martorana/Shutterstock.com; 329 top © Peter Cairns/RSPB Images, bottom © Laurie Campbell; 330 © Andy Hay/RSPB Images; 331 © James Lowen; 332 © Andy Hay/RSPB Images; 333 © James Lowen; 334 © Julian Baker (JB Illustrations); 337 © Stephen Spraggon/Alamy Stock Photo.

POEMS

260 'Survey North of 60 degrees', from Element © Rachel McCarthy (smith/doorstop, 2015); 261 'Utilomar' © Kathy Jetñil-Kijiner.

Index

Page numbers in **bold** refer to illustrations.
Page numbers in *italics* refer to figures and tables.

Abramis brama 283
abundance 31–4
  amphibians 132–5
  beetles 111–13
  birds 150–9
  flies 110–11
  freshwater fish 126–9
  fungi 167–8
  Hemiptera 102–3
  Hymenoptera 107–9
  invertebrates 80–2
  Lepidoptera 87–93
  mammals 143
  Odonata 94–100
  Orthoptera 104–5
  plants 54–67
  reptiles 137–9
  slugs and snails 114–15
  spiders 113–14
  vertebrates 123
Accipiter nisus 182
Acer campestris 43
  pseudoplantus 43
acid grasslands 190, **191**
acid rain 13, 116, 196, 242, 287
acidification 330
  seawater 170, 215–16, 239–42, 291
Acrocephalus scirpaceus 201
Acrocnida brachiata 296
Actaea spicata 62
Actitis hypoleucos 156
Adder 137, 138, **321**
Admiral, Red 79, 83, **83**, 90
Aesculus hippocastanum 41, 43
Aeshna caerulea 99
  cyanea 94
  grandis 96
  mixta 95

  viridis 277
Agabus bipustulatus 112
Agaric, Fly 165, **165**
Agoliinus lapponum 113
agricultural land 207–10
agrochemicals 64, 88, 101, 117
Agrostemma githago 55
Alaria esculenta 218
Alca torda 230–1
Alchemilla glabra 207
Alder **42**, 69, 71
Alectoria nigricans 193
Alga, Frogspawn 171
algae 39, 164, 216, 219, 220
  algal blooms 171–2, **172**, 186
Allen, Myles 249
Alliaria petiolata 43, 208
Alnus glutinosa 43
Alopecurus pratensis 43
Alytes obstetricans 178
Amanita muscaria 165
  ovoidea 167–8
amberjacks 228
Ameletus inopinatus 111
Ammodytes tobianus 222
Amphibian and Reptile Conservation 136
amphibians 14, 19, 25, 119, 120, 130–2, 160, 184, 186, 225
  acid rain 242
  chytrid fungus 177–8
  distribution and abundance 132–5
  migrations 270, 307
  Saprolegnia 178
Amphipyra tragopoginis 93
Amphiura liformis 296
Anarhichas lupus 220
Anas penelope 156

  platyrhynchos 156
Anax imperator 21
  parthenope 98
Anchovy 227, **228**, 235, 291, 297
Andrena avipes 109
  nigroaenea 208
Anemone nemorosa 43
Anemone, Wood 43, **47**, 197
Angelshark 293
Anguis fragilis 137
Anthocharis cardamines 29
Anthoxanthum odoratum 190
Anthus pratensis 201
ants 85, 109
Aphantopus hyperantus 84
aphids 100–2, 104, 109
Apis mellifera 106
Applemoss, Rigid 62
Apus apus 122
Aquatic Coleoptera Conservation Trust 112
Aquila chrysaetos 316
archaeophytes 54–5, 119
Arctia caja 91
Arctostaphylos alpina 57
Ardea alba 152
  cinerea 155
  purpurea 152
Arenaria norvegica 61
Arenicola marina 293
Arenocoris falleni 102
Argiope bruennichi 113
Argus, Brown 88, 91, 117
  Northern Brown 89, **89**
  Scotch 89
Aricia agestis 88
  artaxerxes 89
Armillaria solidipes 167

Index

Arnoglossus laterna 228
Arocatus longiceps 103
Arrhenius, Svante 13
art 258–9
Arthurdendyus triangulatus 81
Arvicola amphibius 324
Ash 43, 50
   ash dieback **176**, 176–7
Ashdown Forest, Sussex **206**
Asio flammeus 322
Asterionella formosa 171, 172
Athysanus argentarius 102
Attenborough, David 263
Austropotamobius pallipes 81
autumn 15, 22, 48, 49–51, 197, 205
   autumn cooling 94
   autumn migrations 120, 124, 144, 148, 158
   fungi 183–6
   lawn-mowing **254**
   lichens 288
Avens, Wood 197
Aythya ferina 212
Azalea, Trailing 60, **60**, 193

bacteria 21, 163, 171, 173, 174
Badger 143, 160, 179, 253
Balaena mysticetus 236
Balanus crenatus 218
Balfour-Browne, Frank 28
Balistes capriscus 228
Baneberry 62
Banks, Brian 96
Barkham, John 159, 286
Barnacle, Montagu's Stellate 294
   Poli's Stellate 220
   Wart 218
barnacles, acorn 218, 220, 293
   cold-water **293**
Barnett, Ray 102
Bassenthwaite, Cumbria 129, **281**, 281–2
Bat, Daubenton's 141
   Greater Horseshoe **140**, 140–1
Batrachochytrium dendrobatidis 177–8
Batrachospermum torfusum 171
bats 140–1, 143, **161**

Bayes, Thomas 26–7
Bear, Polar 335
Bearberry, Alpine 57
Beaver 265
Bedstraw, Heath 190
Bee, Buffish Mining 208–9, **209**
   Dark Blood 109
   Honey 106
   Violet Carpenter 109
   Yellow-legged Mining 109
Beech 43, 50, 71, 196, 200, 327
bees 106–7
   bumblebees 106, 108–9, 319
   distribution and abundance 108–9
Beetle, Great Diving **319**
   Great Silver Water 77, 112
   Lesser Silver Water 28–9
   Northern Dung 113
   Screech 112, **112**
beetles 31, 36, 111–13, 208, 318
Beewolf 107
Ben Lawers National Nature Reserve, Scottish Highlands 58–9, **189**, 190, 192
Betula pendula 43
   pubescens 48
Bib (Pouting) 226
Bifurcaria bifurcata 218
Bilberry 270, **270**
Birch, Downy 48
   Silver 43, 48, **49**, 50, 69
birds 119, 144–9, 160, 229, 307, 308, 324, 326, 329
   birds across the UK 211–13
   caterpillars 35, 146, 181–2, 198–201
   computer modelling 271–2, 286
   distribution and abundance 150–9
   estuaries and mudflats 221
   extreme weather 249
   pathogens 177
   saltmarsh 320–2
   seabirds 120, 215, 222, 224, 230–4, 243, 292, 296, 308, 315–16, 326, 332
   tidal barrages 317–18
   wetland birds 331

   wind turbines 315–16
   woodlands 195, 199–201
BirdTrack 155
Bittern 33, **33**, 271, 333
   Little 34, **34**, 152
Blackbird 122
Blackcap 122, 148
blackflies 100
Blackthorn 43, 102
Blake, William 207
Blenny, Black-faced 228
   Red 228
Blicca bjoerkna 127
Blue, Adonis 87
   Common **36**, 84
   Holly 78, **78**, 83, 84
   Large 85
   Silver-studded **75**, 85
Bluebell **42**, 43, 56, 68, 196
bluetongue virus 179, **179**, 251
Boar, Wild 265
Bohr, Nils 37
Bombus hortorum 106
   hypnorum 108
   lapidarius 79
   monticola 108
Bos taurus 141
Botanical Society of Britain and Ireland (BSBI) 39
Botaurus stellaris 33
bovine tuberculosis (bTB) 178–9
Box 102
Box Hill, Surrey 102
Brachypodium sylvaticum 197
Brachytron pratense 96
Bracken 197, 204, **204**
braconid wasps 208
Bramble 43, **51**
Branta bernicla 321
   leucopsis 149
Bream, Common 283
   Couch's Sea 228
   Gilthead 228
   Silver 127
Breeding Bird Survey (BBS) 155
Brewis, Lady Anna 254–5
Brimstone 29, **76**, 79, 83
Bristol Channel 317–18
British Arachnological Society 113

British Dragonfly Society 94, 95
British Lichen Society 169
British Mycological Society 165
British Trust for Ornithology
    (BTO) 120, 144, 212, 257
  Breeding Bird Atlas 150–1, 256
  Common Birds Census (CBC)
    155
brittle stars **294**, 295, 296
Brome, False 92, 197
Brooks, Steve 95
Bruthwaite Forest, Cumbria **330**
bryophytes 39, 62, 67, 189, 190,
    193, 329
bryozoans 218
Bubulcus ibis 152
Bucephala clangula 331
bud-burst 42, 73, 198, 299
Buddleia 45
Buddleja davidii 45
Bufo bufo 130
    calamita 21
Bug, Box 102, **103**
    Cinnamon 103
    Plane Tree 103
Buglife 31
Buglossidum luteum 226–7
bugs 100–3, 117, 208
Bumblebee, Bilberry 108, **108**
    Garden **16**, 106
    Red-tailed 79
    Tree 108
bumblebees 106, 108–9, 319
Bunting, Snow 194, **194**, 271
Burbot 127, 128–9
Burhinus oedicnemus 271
Bush-cricket, Bog 105
    Grey 105
    Roesel's **16**, 104, **105**
    Southern Oak 105
    Speckled 105
business leaders 263–4
Buteo buteo 123
butterflies 22, 36, 45, 76, 78–9,
    80, 83–5, 194, 208, 249,
    265, 308, 319
  distribution and abundance
    87–91
  emergence 270
  experiments 300, 301–2, 328

niche broadening 100, 105,
    208
  range limits 93, 95, 116, 127,
    331
Butterfly, Wall 93
Butterfly Conservation 86
Buxus sempervirens 102
Buzzard 123, 212, 336

Cabot, David 149
Cacopsylla moscovita 198
Cairngorms 58, 190, **184**, 192,
    192, 289–90, 303
Calanus finmarchicus 225, **225**,
    296–7
    helgolandicus 225, 297
Calidris alba 221
    alpina 156
    canutus 221
    ferruginea 148
Calluna vulgaris 60
Calopteryx splendens 96
Cambridgeshire 28, **145**, **333**,
    335–6
Cameraria ohridella 92–3
Cameron, David 264
Campanula rotundifolia 197
Campylopus flexuosus 193
Canis lupus 210
carabid beetles 36
Carassius carassius 283
carbon dioxide 13, 211, 216,
    239–40, 242, 264, 314, 325,
    329
Cardamine pratensis 43
Cardiff Bay barrage 317–18
Carex bigelowii 60
    echinata 190
    lepidocarpa 190
    nigra 190, 197
    panacea 190
    pilulifera 197
    saxatilis 59
Carney, Mark 263–4
Carp, Common 282, 283, **284**
    Crucian 283
Carpet, Garden 86
Carson, Rachel *Silent Spring* 102
cartoons 262
Castor fiber 265

Catchfly, Alpine 57
caterpillars 35, 146, 181–2,
    198–201
Cattle, Chillingham 141, **141**
Celandine, Lesser **42**, 43
Celastrina argiolus 78
Celloporella hyalina 218
Centaurea cyanus 55
centipedes 75, 77
Centre for Biodiversity Research,
    Trinity College Dublin 24
Centre for Ecology and
    Hydrology (CEH) (UK) 23,
    35–6, 79, 255
Cephalanthera rubra 62
Cerastium nigrescens 61
    tomentosum 69
Cercyon sternalis 112
Certhia familiaris 155
Cervus elaphus 141
cetaceans 222, 235–7, 239
Cettia cetti 150
Chaffinch 212
Chalcolestes viridis 98
Channel Islands 167
Character, Hebrew 86
Charadrius dubius 212
    hiaticula 221
Charr, Arctic 129, **129**, 186
Chaser, Broad-bodied 96
Cheshire 12, 28–9
Chicory 55, **55**
Chiffchaff 122, 151
Chinnor, Oxfordshire 44
Chirocephalus diphanus 326
Chloris chloris 145
Chorosoma schillingi 102
Chorthippus albomarginatus 105
Chthamalus montagui 294
    stellatus 220
Chub 283, **283**
chytridiomycosis 177–8
Cichorium intybus 55
Cinclus cinclus 145
Circus cyaneus 156
Cladonia coccifera 193
Clarke, Harry 85
Clegg, John *The Freshwater Life of
    the British Isles* 171
climate change 11, 37

evidence for change 11–17
exploring the evidence for change 22–35
monitoring climate change in the UK 35–7
what change might mean for wildlife 17–21
Climate Change Act 2008 (UK) 264
climate change defences 311
 flood protection 322–4
 renewable energy 311–19
 sea-level rise 109, 320–2
climate envelopes 17–18, 20, 68, 182, 308, 326, 328, 331
 computer modelling 277–8, 280, 287, 290, 292, 294–5, 297, 303–4
 insects 89–90, 108, 112
 plants 207, 253, 270, 272, 274
Cloudberry 18–19, **19**
Club of Rome *The Limits to Growth* 305
Clubmoss, Fir 60, 193
clubmosses 39
Clupea harengus 235
Clutton-Brock, Tim 142
coastal environments 215–16, 242–3
 coastal kelp forests 218–19
 coastline 216–18
 estuaries and mudflats 221
 out at sea 222–4
 rocky shores 219–21
Coccinella 7-punctata 79
Cock's-foot 43
Cod, Atlantic 226, **229**, 290
Coenagrion hastulatum 99
 mercuriale 279
Coenonympha tullia 89
Colias croceus 90
Cololejeunea minutissima 67
Colt's-foot 43
Colura calyptrifolia 67
Combes, Jean 23
Comma 79, **88**, 91, 117, 256
communities 34–5, 181–3, 213
 agricultural land 207–10
 birds across the UK 211–13
 dunes and heaths 202–6

freshwater systems 184–8
mountains and moorlands 188–94
rewilding 210–11
woodlands 195–201
computer modelling 268–72
 invertebrate models 277–80
 lichen models 287–90
 new horizons and limits to predictive methods 304–5
 North Sea models 290–7
 plant models 272–7
 vertebrate models 281–7
Conchological Society of Great Britain 114
Conehead, Long-winged 16, **104**, 104–5
 Short-winged 105
Confederation of British Industry (CBI) 263
conifers 39, 167, 329, 330
Conocephalus discolor 104–5
 dorsalis 105
conservation 307
 coping with climate change defences 311–24
 future prospects 334–6
 managing for climate change 325–33
 where are we now? 307–10
copepods 225, 296–7
Corbett, Keith 136
Coregonus albula 129
 lavaretus 129
Corizus hyoscyami 103
Corncockle 55
Cornflower 55, **55**
Cornu aspersum 78
Coronella austriaca 138
correlation 70–3
Corvus corone 212
Corylus avellana 43
Coturnix coturnix 271
Country Stewardship 87, 327, 335
Coventry University 255
Cowberry 193
Cowlwort, Fingered 67
Crab, Chinese Mitten 81
Crane, Common 331, **331**, 333

craneflies **329**
Cranesbill, Wood 207
Crassostrea gigas 221
Crataegus monogyna 43
Crayfish, Signal 81
 White-clawed 81
Crepis biennis 65
Cricket, Mole 82, **82**
crickets 104–5
Cridland, John 263
Crocothemis erythraea 98–9
Crocus vernus 47
Crocus, Spring 47
Crow, Carrion 212, **212**
crustaceans 77, 219, 222
Cuckoo 22, 62, 121, 122, 151, **151**, 201, **201**
Cuckooflower 43
Cuculus canorus 22, 62, 122
Culicoides imicola 179
 obsoletus 179
 pulicaris 179
culture of climate change 258
 art 258–9
 cartoons 262
 films 262
 poetry 260–1
Cumbria 18, 87, 108, 114, 129, 171–2, 186, 281, 330
Curlew 159, 221
Curtain Crust, Yellowing 167, **167**
Cyanistes caeruleus 122
cyanobacteria 172
Cygnus cygnus 331
Cylotella 171, 172
Cymbalaria muralis 45
Cyprinus carpio 282

Dabberlocks 218
Dactylis glomerata 43
Dactylorhiza praetermissa 65
Daffodil, Sea 56, **56**
Daisy, Oxeye 43
Daltonia splachnoides 67
Daltonia, Irish 67
Dama dama 119
damselflies 76, 93–4
 distribution and abundance 94–100

Damselfly, Large Red 94
  Northern 99
  Small Red-eyed 98, **98**
  Southern 279–80, **280**
  Southern Emerald 98
  White-legged 98, 100
  Willow Emerald 98
Danaus plexippus 80
Dandelion 54, **54**
Danthonia decumbens 197
Darter, Banded 99
  Red-veined 99
  Ruddy 95, **95**, 96
  Scarlet 98–9
  White-faced 99, **99**
Dartmoor **246**
Darwin, Charles 14
Dead-nettle, White 45, **45**, 55
deer 175, 176, 196
Deer, Fallow 119
  Red 141–2, **142**, 179
  Reeves Muntjac 196
Deergrass 59, 190
Delesseria sanguinea 218
Delichon urbicum 100, 122
Delphinapterus leucas 226
Delphinus delphis 235
Demoiselle, Banded 96, 98
demonstrations 11, 17
Denton, Jonty 16
Dermochelys coriacea 237
Derwentwater, Cumbria 129
Deschampsia cespitosa 197
  exuosa 203
Devon 114–15, 132, 159
Dichodontium palustre 59
Dicranella heteromalla 193
Dicranum fuscescens 59
  scoparium 190
Dipper 145
Dipturus batis 293
Distomus variolosus 218
distribution 27–31
  amphibians 132–5
  beetles 111–13
  birds 150–9
    Comparing hypothetical records to detect real change 29
  flies 110–11

freshwater fish 126–9
fungi 167–8
Hemiptera 102–3
Hymenoptera 107–9
invertebrates 80–2
Lepidoptera 87–93
mammals 143
Odonata 94–100
Orthoptera 104–5
plants 54–67
reptiles 137–9
slugs and snails 114–15
spiders 113–14
vertebrates 123
DNA 71, 163, 174
Dog-violet, Common 40, 197, **197**
  Early 197
Dogfish, Spiny (Spurdog) 228
Dolichovespula media 108
  saxonica 108
Dolphin, Fraser's 235
  Short-beaked Common 235, **235**
  Striped 235
  White-beaked Dolphin 236
dormice 140
Dorset 55, 130, 133, 136, 138, 152, 204, 225, 228
Dove, Turtle 122, 150, **150**
Draba incana 62
dragonflies 76, 93–4
  distribution and abundance 94–100
Dragonfly, Emperor 21, 94, **96**, 96–7
  Hairy 96
  Lesser Emperor 98
  Northern Emerald 99
drainage ditches 329, **329**
dredging 332–4, **324**
Dreissena polymorpha 81
drones 53
Dumfries and Galloway 139
dunes 202–6
Dunlin 156, 159, 221
Dunnock 201
Durham 96, 108
Durham University 90
Dusty Miller 69

Dysstroma truncata 86
Dytiscus 77
  circumflexus 112
  marginalis **112**

Eagle, Golden **307**, 316
earthworms 81, 143, 158
EC Directive on the Conservation of Wild Birds 318
Echiichthys vipera 228
Echinocardium cordatum 296
ecosystems 181
ectotherms 20–2, 75, 119, 120, 178
Eel, European **314**
Egret, Cattle 152
  Great White **119**, 152
  Little 152, **152**, 166
Egretta garzetta 152
Ehrlich, Paul and Anne *The Population Bomb* 305
Eiseley, Loren 11
Elder 43
Electra pilosa 218
Emerson, Ralph Waldo 298
endotherms 20, 119, 120, 140, 154–5, 237
England 14–16, 20, 25, 184, 192, 250, 253, 309
  amphibians 131–2
  birds 34, 100–1, **119**, 123, 147, 149, 151–4, 158, 213
  coastline 218, 220, 320
  conservation 319, 320, 329
  fish 127–9, **227**, 228, 282
  fungi 167, 170, 177
  insects **82**, **85**, 87–8, 90, 92–3, 95–9, 102, 105–14, 277–8
  lichens 288
  mammals 140, 143
  plants 62, 64–6, 330
  reptiles 136, 138
  slugs and snails 114–15
  woodlands 195, 196
English Channel 37, 92, 103, 109, 137, 143, 218, 220–1, 224, 228, 294

Index

Engraulis encrasicolus 227
enhanced vegetation index (EVI) 51, 52–3
Ennomos erosaria 93
Enochrus melanocephalus 112
Entelurus aequoreus 233
Environmental Change Institute, Oxford University 36
Environmental Change Network (ECN) 35, 58
Ephemera danica 111
Epidalea calamita 21
Epipactis phyllanthes 65
Epistrophe melanostoma 110
Erebia aethiops 89
   epiphron 300
   medusa 300
Erica ciliaris 40
   cinerea 197
Erinaceus europaeus 140
Eriocheir sinensis 81
Erithacus rubecula 120, 122
Erynnis tages 84
Erythromma viridulum 98
Esox lucius 129
Essex 98, 105, 111, 217, 320, 333
estuaries 221
EU Habitats Directive 318
Eupalagia quadripunctaria 91
European Union 335
eutrophication 35, 172, 183, 184–6, 188, 190, 192, 203–4, 213, 301, 325, 330
Everett, Sue 196
evidence for change 11–17, 22
   communities 34–5
   phenology 22–7
   species abundance 31–4
   species distributions 27–31
experiments 298
   invertebrate experiments 300–2
   lichen experiments 303
   plant experiments 298–9
   vertebrate experiments 302–3
extreme weather 14, 17, 106, 204, 216, 232, 245–9, 293, 322

Fagus sylvatica 43
Fairy Ring Champignon 166, **166**
Falco tinnunculus 155
farming 250–3
   agricultural intensification 55, 88, 213, 325
   agricultural land 207–10
   intensive farming 91, 101, 151–2, 253
Feather-moss, Red-stemmed 59
ferns 39
fertilisers 91, 207, 250
   eutrophication 184
Ficedula hypoleuca 146
Fieldfare 122
films 262
fish 119
   distribution and abundance 126–9
   fish passes **313**
   freshwater fish 124–9
   marine fish 226–30
   river barriers 312–14
Fitter, Richard 44–7, 68–9
FitzRoy, Robert 14, 22
Flatworm, New Zealand 81
Flavocetraria nivalis **289**, 303
Flavoparmelia soredians 170, **170**
flies 110–11
floods **16**, 248, 248–9, **323**
   flood protection 322–4
Flycatcher, Collared 196
   Pied 146, **147**, 159
   Spotted 122
Fork-moss, Broom 190
   Dusky 59, 60, 193
Forklet-moss, Marsh 59
   Silky 193
fossil fuels 13, 242, 305, 311–12, 314, 334
Foster, Garth 112
Fox, Red 160
Foxtail, Meadow 43
Fratercula arctica 231
Fraxinus excelsior 43
Freshwater Biological Association (FBA) 124
Freshwater Habitats Trust 98
freshwater systems 184–8

Friends of the Earth 263
Fringilla coelebs 212
Fritillary, Glanville 331
Frog, Common 24, **121**, 122, 130, 178, 186–8
   Edible **134**, 134–5
   Marsh 134
   Pool 130–1, 134
frogs 23
   spawning times 24–5, 40, 130–2, 134–5, 186–7
Frostwort, White 59
fungi **163**, 163–4, 179
   distribution and abundance 167–8
   phenology 164–7
Fungus, White Egg (Bearded) **167**, 167–8
fungus gnats 208
future developments 267
   computer modelling 268–97
   experiments 298–303
   new horizons and limits to predictive methods 304–5

Gadus morhua 226
Gagea serotina 58
Galanthus nivalis 40, 43
Galapagos Islands 222, 224
Galium saxatile 190
Gallinago gallinago 212
Gange, Tim 165
Gannet **315**, 315–16
gardening 39, 42, 253–5
Gasterosteus aculeatus 126
gastropods 21, 77
Gatekeeper 84
Geltsdale, Cumbria 330
genetic studies 70–1
Gentian, Alpine 57
Gentiana nivalis 57
Geranium sylvaticum 207
geraniums 88
Geum urbanum 197
Gibbula cineraria 218–19
   umbilicalis 220
glaciation 11, 13, 186
Glastonbury Tor, Somerset **336**
Gleed-Owen, Chris 139
global warming 13–15

359

standard climate models
(HadCM3) 268–72
Gloucestershire 140, 196
Gobio gobio 283
Godwit, Bar-tailed 221
Goldeneye 331
Goldenfleece, Prickly 254
Gonepteryx rhamni 29
Gonocerus acuteangulatus 102
Goosander 150
Goose, Barnacle 149, **149**
  Brent 321
Gore, Al 262, **262**, 264, 325
grape vines **253**
Grasshopper, Lesser Marsh 105
  Stripe-winged 105
  Woodland 105
grasshoppers 104–5
Grayling (butterfly) 84
Grayling (fish) 127
Great Fen 210, **211**
Greenfinch 145, 177
greenflies 100, **101**
greenhouse gases 13, 239, 243, 250, 264, 282, 296
  cuts in emissions 310, 336
  lichens 287–8
  seafloor species 294–5
Greenpeace 263
Griffiths, Richard 133–4
ground-truthing 51
Groundhopper, Slender 105
Grouse, Black 150, 156, **156**
Grus grus 331
Gryllotalpa gryllotalpa 82
Gudgeon 283
Guernsey 48
Guillemot 230–1, **231**
Gulf Stream 56, 222–4, 269–70
Gull, Mediterranean 331
Gymnadenia conopsea 31
Gymnocephalus cernua 127
Gymnomitrion concinnatum 59
Gyomitra esculenta 166
Gyrinus urinator 112

habitats 20, 21, 181–3
  agricultural land 207–10
  birds across the UK 211–13
  dunes and heaths 202–6

freshwater systems 184–8
mountains and moorlands 188–94
rewilding 210–11
woodlands 195–201
HadCM3 268, 272
Haematopus ostralegus 73
Hair-grass, Tufted 197
  Wavy 203
Halibut 226, 229
Halichoerus grypus 237
Hampshire 16, 108, 131, 255
Hare, Brown 119, 319, 336
  Mountain 194, **194**
Harebell 197, **197**
Harmonia axyridis 111–12
Harrier, Hen 156
Hawk's-beard, Rough 65
Hawker, Azure 99
  Brown 96
  Green 277–9, **278**
  Migrant 95, 96
  Southern 94, 95, **95**
Hawthorn 43, 50, 102
hay meadows 55, 207
Hazel 43, 50
Heath, Blue 57, **57**
  Dorset 40
Heath, Large 89
Heath-grass 197
Heather, Bell 197
  Ling 60, 193, **193**
heaths 202–6
Hebrides 112, 138, 328
  Lewis 316
  Rum 141–2
Hedera helix 43
Hedgehog 140, 336
Helleborine, Green-flowered 65
  Red 62
Hemiptera 100–2
  distribution and abundance 102–3
Henbane, White 254
Henosepilachna argus 111
hermit crabs 295, **295**
Heron, Grey 155
  Purple 152, **152**
Herring 235, 290
Hesperia comma 87

Hewitt, Stephen 30
Himantoglossum hircinum 62
Hindley, Tim 165
Hipparchia semele 84
Hippoglossus hippoglossus 226
Hippophae rhamnoides 204
Hirundo rustica 42, 122
Holcus lanatus 43
Holly 43, 87
Holocene 12
honey fungus 167
Honeysuckle 106, 197, 325
Hook-moss, Curled 190
hops 88
Hornet, European 107, **107**
hornworts 39
Horse-chestnut **41**, 41, 43, 50, 93
horsetails 39
Hoverfly, Hornet Mimic 110, **110**
Hudson, W, H. *Hampshire Days* 184
Humber 320
Humboldt Current 222
Huntley, Brian 90
Huperzia selago 60
Hutchings, Mike 31
Hyacinthoides hispanica 56
  non-scripta 43
hybridisation 56–7
Hydrochara caraboides 28–9
hydroelectric power 312–14
Hydroid, Knotted-thread 218
Hydrophilus piceus 77
Hydroporus scalesianus 112
Hydrovatus cuspidatus 112
Hygrobia hermanni 112
Hygromia cinctella 114
Hygrotus nigrolineatus 112
Hylocomium splendens 190
Hymenoptera 106–7
  distribution and abundance 107–9
Hymenoscyphus fraxineus 176–7
Hyoscyamus albus 254

ice ages 11–12, 305
  Ice Age 13, 126, 182
  Little Ice Age **12**, 13, 152

Index

Ichthyosaura alpestris 130
Ilex aquifolium 43
Ilybius chalconatus 112
Inconvenient Sequel, An: Truth to Power 262
Inconvenient Truth, An 262
insects 19, 75–7, 84–5
  insect samplers **85**
  Rothamsted Insect Survey (RIS) 86, 91, 101–2
Intergovernmental Panel on Climate Change (IPCC) 264
International Council for the Exploration of the Sea 226
invertebrates 75–7, 116–17
  computer modelling 277–80
  distribution and abundance 80–2
  flies and beetles 110–13
  Hemiptera 100–3
  Hymenoptera 106–9
  invertebrate experiments 300–2
  Invertebrate phenology measures in the UK 79
  Lepidoptera 83–93
  Odonata 93–100
  Orthoptera 104–5
  phenology 77–9
  slugs and snails 114–15
  spiders 113–14
Ireland 24, 39, 81
  birds 101, 149, **152**, 158
  fish 124, 228
  insects 86, 96, 112
  mammals 143, 235
  plants 67, 205–6
  slugs and snails 114–15
Irfon River, Wales **241**
Irish Sea 228
iron bacteria **173**
Isle of Man 88, 112, 176
Isle of Mull 59, 190
Isle of Wight 168, 221
Ivy 43, 87
Ixobrychus minutus 34
Ixodes ricinus 175–6

Jefferies, Richard *Wild Life in a Southern County* 40

jellyfish 237–9
Jetñil-Kijiner, Kathy *Utilomar* 261
Joint Nature Conservation Committee Marine Biodiversity Monitoring programme 232
jumping plant lice 198
Juncus effusus 190
  squarrosus 190
Jynx torquilla 271

Keats, John 50
Keen of Hamar, Shetland **61**, 61–2
kelp 216
  coastal kelp forests 218–19
Kent 64, 152, 246
Kerr, Richard 14, 108
Kestrel 155, 212
Kite, Red 123, 336
Kittiwake 231, 234, **234**, 243, 292, 296–7, 332
Knot 221
Knowler, John 92
Kogia sima 235
Koppett, Leonard 304

laboratory studies 71–3
Laccobius simulatrix 112
Lacerta agilis 32
Lactuca serriola 64
  virosa 64
Lady Park Wood, Gloucestershire 196
Lady's-mantle, Smooth 207, **207**
Ladybird, Bryony 111, **111**
  Harlequin 111–12
  Seven-spot 79
Lagenodelphis hosei 235
Lagenorhynchus albirostris 236
Lagopus muta 194
Lakenheath Fen 333
Laminaria hyperborea 218–19
  ochroleuca 218–19
Lamium album 45
Lancashire 28–9, 109, 184, 245
Lapwing 155, 156, 322
Larch 43
Larix decidua 43

Larus melanocephalus 331
Lasius neglectus 109
Lassiommata megera 93
leaf-fall 44, 49, **50**
Leaf Miner, Horse-chestnut 92–3, 177
leaf tinting 42–4
leafhoppers 102, 208
Lecanora populicola 288
leeches 77
Lepidoptera 83–6, 93
  distribution and abundance 87–93
Lepidorhombus whiffiagonis 229
Leptophyes punctatissima 105
Leptothrix ochracea 173
Lepus europaeus 119
  timidus 194
Lestes barbarus 98
Lettuce, Great 64
  Prickly 64, 272–4, **273**
Leucanthemum vulgare 43
Leucoraja circularis 293
  naevus 229
Leucorrhinia dubia 99
Libellula depressa 96
Lichen, Madame's Pixie-cup 192
  Whiteworm 60, 193, **193**
  Witch's Hair 193
lichens 163, **169**, 169–70, 179
  computer modelling 287–90
  lichen experiments 303
Lilac 43
Lily, Snowdon 58, **58**
Limacus flavus 115, **115**
  maculatus 115
Lime, Small-leaved 18, **18**, 69
Limnebius crinifer 112
Limosa lapponica 221
Limpet, Black-footed 220
  Blue-rayed 219
  Common 220, **220**
limpets 216, 294
Lincolnshire 87, 113
Ling 228
Liparis loeselii 62
Lissotriton helveticus 122
  vulgaris 122
Little Ice Age **12**, 13, 152
liverworts 39, 59, 62

Lizard, Sand 32, **32**, 135–9, 161, 176, 206, 328
  Viviparous **137**, 137–8, 161, 176
  Wall 139
Llyn Peninsula, Wales 228
Loiseleuria procumbens 60
Lonicera periclymenum 106
Lophius piscatorius 229
Lorenz, Edward 304
Lota lota 127
Lugworm 293
Luscinia megarhynchos 122
Lutra lutra 323
Lychnis alpina 58
Lyme disease 175–6
Lynx 210, 265
Lynx lynx 210

Mabey, Richard 73
Mackerel, Atlantic 291
  Atlantic Horse 291, **291**
maize 253, **253**
malaria 175
Mallard 156
mammals 140–2
  distribution and abundance 143
  marine mammals 235–7
mammoths 182
managing for climate change 325
  general strategies 326–8
  grand plans 332–3
  ideas and actions in the uplands 328–30
  individual contributions 325
  managing the marine environment 332
  options to assist climate change beneficiaries 331
Manning, Mary 23
Maple, Field 43, 50
Marasmius oreades 166
Marbled Carpet, Common 86
marine environments 215–16, 222–6, 242–3
  changes in seawater acidity 239–42
  changing fish faunas 226–30
  computer modelling 290–7

fates of seabirds 230–4
managing the marine environment 332
marine mammals and reptiles 235–9
Marine Protected Areas (MPAs) 332
Marren, Peter 164, 165
Marsh-orchid, Southern 65
Marshall Islands, Pacific 261
Marsham, Robert 23, 79, **245**, 255
Marten, Pine 265
Martes martes 265
Martin, House 100, 109, 122, 149, 151–2, **152**, 157
  Sand 122, **156**, 157
Mat-grass 59, **59**, 190, 328
Maunder Minimum 12
May, Robert 75
Mayfly, Green Drake 111, **301**
  Upland Summer 111
McCarthy, Rachel *Survey north of 60 degrees* 260, **260**
Meconema meridionale 105
Medicago arabica 65
Medick, Spotted 65, **66**
Megrim 229
Melanargia galathea 87
Meles meles 143
Melitaea cinxia 331
Meloe 31
Membranoptera alata 218
Mergus merganser 150
MERIS Terrestrial Chlorophyll Index (MTCI) 51
Mersey 184, 204, 328
Mesoplodon densirostris 235
Meteorological Office 23
  Hadley Centre (HadCET) 14, 268–9, **269**
Metrioptera brachyptera 105
  roeselii 104
microbes 163, 171–3, 179
Microsphaera alphitoides 168
  azaleae 168
midges 179, 188
mildews 168, **168**
millipedes 75, 77
Milvus milvus 123

Mistletoe 20, **20**
Molinia caerulea 105
molluscs 115, 219
Molva molva 228
Monarch 80
MONARCH (Modelling Natural Resource Responses to Climate Change) 271, 279
Monbiot, George 263
money spiders **115**
monitoring climate change in the UK 35–7
Monkfish, Angler 229
Monodon monoceros 236
Moor-grass, Purple 105
moorlands 188–94
Morchella esculenta 164
Morel, False 166, **166**
morels 164, **165**
Morus bassanus 314
Moss, Alpine Haircap 190
  Pohlia 193
  Racomitrium 193
  Rusty Swan-neck 193
Moss, Brian 67, 184
mosses 39, 190, 193
Moth, Mouse 93
  Winter 198–9, **199**
moths 85–6, 93, 208
  distribution and abundance 91–3
  micro-moths 85
mould beetles 208
mountains 188–94
  satellite technology 52–3
  subarctic alpine plants 57–62
Mouse-ear, Shetland 61
mudflats 221
Mullet, Red 227, **227**
Mullus surmuletus 227
Muntiacus reevesi 196
Muscicapa striata 122
Mussel, California 241
  Zebra 81
Mustard, Garlic 43, 208
Mustela putorius 123
Mycobacterium bovis 178
Mylia anomala 62
Myotis daubentonii 141
Mytilus californianus 241

Nardus stricta 59
Narwhal 236, **236**
National Biodiversity Network (NBN) 27, 31, 96
National Moth Recording Scheme 86
Natrix helvetica 139
Natural Environment Research Council (NERC) *Climate Change and Biodiversity Report Card* 37
naturalists 255–7
Nature Conservancy Council 325, 329
Nature Conservation Marine Protected Areas (MPAs) 332
Nature's Calendar 39, 44, 49, 77, 106, 121, 146, 255
  Invertebrate phenology measures in the UK 79
  Plant phenology measures in the UK 43
  Vertebrate phenology measures in the UK 122
Naucoris maculatus 103
Nebrioporus canaliculatus 112
Neodenticula seminae 226
neonicotinoids 102
neophytes 55, 57, 69, 119
Nettle, Stinging 87, 88
New Forest 82, 107, **160**, 165, **195**, 196
New York Times **262**
Newcastle upon Tyne 234
Newt, Alpine 130
  Great Crested 30, 122, 130, **133**, 133–4, 268
  Palmate 122, 130
  Smooth 122, 130
newts 15
  predation of frog tadpoles 187–8
Nightingale 122
non-government organisations (NGOs) 31, 32, 263, 325
Norfolk 23, 79, 81, 115, 333
normalised difference vegetation index (NDVI) 51, 52–3
North Atlantic Conveyor 269
North Atlantic Oscillation (NAO) 101, 230, 292
North Sea 12, 77, 228–9
  computer modelling 290–7
Northampton 109
Northumberland 154, 265
Nostoc 173
Noterus clavicornis 112
nuclear power 311–12
Numenius arquata 221
Nymphalis c-album 79
  io 79
  polychloros 81–2
  urticae 79

Oak, Pedunculate 43, 50, 71
  Sessile 43, 50, 71, 72, **72**, 299
Obelia genticulata 218
Ochrolechia frigida 60
Ochthebius alpinus 112
Odonata 76, 93–4
  distribution and abundance 94–100
Odontoscelis lineola 102
Oenanthe oenanthe 122
oil beetles 31
Olive, Blue-winged 301
Omocestus rufipes 105
Operophtera brumata 198–9
Ophiothrix fragilis 295
Ophrys sphegodes 208
  tenthredinifera 55
Orange Tip 29, **77**, 78, 83, 89, 208
orchards **251**
Orchid, Fen 62
  Fragrant 31
  Lady 64, **65**, 276–7
  Lizard 62, **63**, 64
  Man 64, **65**
  Sawfly 55, **56**
  Tongue 56
Orchis anthropophora 64
  purpurea 64
Orkneys 59, 190
Orthetrum cancellatum 95
  coerulescens 94
Orthoptera 104–5
Orthosia cerasi 86
  gothica 86
Oryctolagus cuniculus 119

Osilinus lineatus 220
Otter 323
Ouse 324
Ouse Fen, Cambridgeshire **333**
Ouzel, Ring 158, **158**, 286
overpopulation 305, 325, 336
Owl, Barn 155, **155**
  Short-eared 322, **322**
Oxalis acetosella 197
Oxford University 36, 249
Oxford, Geoff 113
Oxfordshire 44–6, 143, 199
Oyster, Pacific 221
Oystercatcher 73, 217, **217**

Pacifastacus leniusculus 81
Packham, Chris 146
Pagrus pagrus 228
Pagurus pridaeux 295
Painted Lady 80, **80**
Palomena prasina 102
Palustriella commutata 190
Pancratium maritimum 56
Papaver rhoeas 64
Papilio machaon 29
Parablennius ruber 228
Pararge aegeria 79
Paris Climate Change Summit 2015 310, 336
Parus major 122
Patella depressa 220
  pellucida 219
  vulgata 220
pathogens 175–9
Peacock 79, 83, 88
Pearce-Higgins, James 146, 155, 257, **257**, 304
Pelophylax esculentus 134–5
  lessonae 130–1
  ridibundus 134
Pelto, Jill 258
  'Landscape of Change' **258**
Pennant, Thomas 238
Perca fluviatilis 124
perceptions of climate change 245, 265
  business leaders and politicians 263–4
  culture of climate change 258–62

extreme weather 245–9
naturalists and scientists 255–7
non-government organisations (NGOs) 263
working outside 250–5
Perch 124, **125**, 186
Perenniporia ochroleuca 167
Periclimenes sagittifer 225–6
Periparus ater 155
pesticides 13, 91, 102, 123, 207–8, 250, 256, 305
Phalacrocorax aristotelis 230–1
Phasianus colchicus 196
Pheasant 196
Phellinus wahlbergii 167
Phengaris arion 85
phenology 22–7
  amphibians 130–2
  birds 144–9
  freshwater fish 124–5
  fungi 164–7
  Hemiptera 100–2
  Hymenoptera 106–7
  invertebrates 77–9
  Lepidoptera 83–6
  mammals 140–2
  Odonata 93–4
  plants 42–53
  reptiles 135–7
  Vertebrate phenology measures in the UK *122*
  vertebrates 121
Philanthus triangulum 107
Philonotis rigida 62
Phleum pratense 43
Phoca vitulina 237
Phocoena phocoena 236–7
Phoenicurus phoenicurus 145
Pholcus phalangioides 114
Phragmites australis 33
Phycodrys rubens 218
Phyllodoce caerulea 57
Phylloscopus collybita 122
  trochilus 122
Pieris brassicae 29
  napi 78
  rapae 78
Pike **125**, 129, 186, 283
Pincushion, Balding 67

Pine, Scots 19, **19**, 69, 330
Pinus sylvestris 19
Pipefish, Snake 233
Pipistrelle, Nathusius' 143, 285–6
Pipistrellus nathusii 143
Pipit, Meadow 201
Plaice, European 290
plant distribution and abundance 54–7
  losers 57–63
  winners 63–7
plant phenology 42–4
  early springs 44–9
  new approaches to investigate phenology 51–3
  Plant phenology measures in the UK *43*
  summers and autumns 49–51
planthoppers 208
plants 39–41, 73
  boreal species 62–3
  computer modelling 272–7
  distribution and abundance 54–67
  how are climate change responses achieved? 68–70
  mechanisms of change 68–73
  moving beyond correlation 70–3
  phenology 42–53
  plant experiments 298–9
  subarctic alpines 57–62
Platycleis albopunctata 105
Platycnemis pennipes 98
Platypalpus aliterolamellatus 30
Plebejus argus 85
Plectrophenax nivalis 194
Pleistocene 11, 226
Pleurococcus 171
Pleuronectes platessa 290
Pleurozium schreberi 59
Plover, Golden 156, 329
  Grey 221
  Little Ringed 212
  Ringed 221
Pluvialis apricaria 156
  squatarola 221
Pochard 212
Podarcis muralis 139

poetry 260–1
Pohlia nutans 193
Polecat 123, 265
politicians 263–4
pollen 41, 48–9, 69, 71, 106
pollinators 48, 109, 181, 208–10, 325
Polyommatus bellargus 87
  icarus 84
Polytrichastrum alpinum 190
Pomatias elegans 21
Poppy, Common 64
Population Matters 336
population, human 305, 325, 336
Porpoise, Harbour 236–7
Pouncewort, Minute 67
Pout, Norway 228
Powan 129, 302
powdery mildews 168, **168**
Prawn, Anemone 225–6
protists 163
Prunella modularis 201
Prunella vulgaris 197
Prunus spinosa 43
Ptarmigan 194
Pteridium aquilinum 197
Puccinia graminis 168
Puffin 231, **233**
Pulcherricium caeruleum 168
Pungitius pungitius 127
Pyrenula macrospora 289
Pyronia tithorus 84
Pyrrhosoma nymphula 94

Quail 271
Quaker, Common 86
Quercus petraea 43
  robur 43

Rabbit 119
rabies 175
Racomitrium 191
  heterostichum 193
rainfall 15–17, 20, 25, 49–50
Ramsar sites 318
Rana temporaria 24, 122
Ransome, Roger 140
Ranunculus ficaria 43
Rat, Brown 175

Rattus norvegicus 175
Ray, Cuckoo 229
  Sand 293
Razorbill 230–1
Reading, Chris 133
Redshank 154, 156, 221, 322, 324
Redstart 145, **145**
Redwing 122
Reed, Common 33, **33**
regression 25–6, 46, 70
  Regression of flowering dates against year 26
renewable energy 311–19
reptiles 135–7
  distribution and abundance 137–9
  marine reptiles 237–9
rewilding 210–11
Rhantus grapii 112
  suturalis 112
Rhinolophus ferrumequinum 140–1
Rhododendron 168, 254–5
rhopalids 102
Ringlet 84
  Mountain 300
  Woodland 300
Riparia riparia 122
Rissa tridactyla 231
Roach 186, 282
Robin 120, 122, 146
rocky shores **216**, 219–21
Rosa canina 43
Rose, Dog 43
Rothamsted Insect Survey (RIS) 86, 91, 101–2
Rothamsted Research Centre, Hertfordshire 101
rove beetles 208
Rowan 43, 50, 71, **71**
Royal Horticultural Society (RHS) 254
Royal Meteorological Society 132
Royal Society 256
Royal Society for the Protection of Birds (RSPB) 121, 263, 265, 316, 330
  Big Garden Birdwatch 144

Rubus chamaemorus 18–19
  fruticosus 43
Rudd 282
Ruffe 127
Rush, Common 190
  Heath 190
rusts 168
Rutilus rutilus 186

Salix herbacea 60
  repens 198
Salmo salar 124
  trutta 128
Salmon, Atlantic 124, **312**
saltmarshes 109, 132, 215, 216–18, 242–3, 320
Salvelinus alpinus 129
Sambucus nigra 43
Sand Eel, Lesser 222
Sander lucioperca 126
Sanderling 221
Sandpiper, Common 156
  Curlew 148, **148**
Sandwort, Arctic 61, **61**
Saprolegnia 178
Sardina pilchardus 226–7
Sardine 226–7, **227**
satellite technology 51–3
  Changing phenology across Scandinavia 52
Savage, Maxwell 132
Saxifraga stellaris 193
Saxifrage, Starry 193, **193**
Scaldfish 228
Scardinius erythrophthalmus 228
Science 14
scientists 255–7
Scillies 80
Sciurus carolinensis 177
  vulgaris 177
Scomber scombrus 291
Scotland 12, 15–16, 19, 25, 52, 210
  amphibians 132–3
  birds 101, 145, 151–2, 156, 158, 213, 231, 296–7
  coastline 217–18, 220, 224
  conservation 316, 329
  fish 129, 186, 227, 282, 291
  fungi 168

insects 86, 88–9, 92, **95**, 96, 99, 105, 108, 110, 112–14, 300
lichens 170, 288
mammals 140, 235, 285
plants 57–62, 67, 189–93, 205, 273, 274–5, 298
reptiles 137, 139
wind turbines 316
woodlands 195, 330
scrub-bashing **309**
Sea Potato 296
Sea Squirt, Lesser Gooseberry 218
Sea-buckthorn 204
sea-level rise 215–18
  coastal defences 109, 320–2
sea trout 313, 314
sea walls **321**
seabirds 120, 215, 222, 224, 292, 332
  fates of seabirds 230–4
  seabird colonies 243, 296–7, 308, 326
  wind turbines 315–16
Seal, Common (Harbour) 237
  Grey 237
seasons 15–17
  early springs 44–9
  phenology 22–7
  summers and autumns 49–51
seawater acidity 170, 215–16, 239–42, 291
seaweeds 216
Second World War 106, 207, 325
Sedge, Carnation 190, 193
  Common 190, 197
  Pill 197
  Rock 59
  Star 190
  Stiff 60, 193
Segestria orentina 114
Selfheal 197
Semibalanus balanoides 220, 293
Serapias lingua 56
Seriola 228
Serratella ignita 301
Severn 221, 317–18, 320
Shag 230–1, **232**, 234

Shelduck 317, **317**
Shetlands 59, 61, 190, 231
Shieldbug, Green 102
shieldbugs 102
Shrimp, Fairy 326
    Tadpole 203, 326
Sigara iactans 103
    longipalpis 103
silken fungus beetles 208
Siskin 212
Sites of Special Scientific Interest (SSSIs) 318, 326
Skate, Common 293
Skimmer, Black-tailed 95, 96
    Keeled 94
Skipper, Dingy 84, 92
    Silver-spotted 87, 331
    Small 87, 90, 256, **302**, 328
Slow-worm 137
Slug, Green Cellar 115
    Irish Yellow 115
slugs 75, 114–15
Smith, Malcolm 24
Snail, Garden 78, 115
    Girdled 114
    Round-mouthed 21
snails 75, 114–15
Snake, Grass 139
    Smooth 138
Snipe 212, 324
snow drifts **246**
Snowdrop 22, 26, 40, **41**, 43, 47, 69
solar output 13, 305
solar power 318–19
    solar farms 318–19, **319**
Sole, Dover 229, 291
    Yellow (Solenette) 226–7
Solea solea 229
Solent 105
Solway 133, 217–18
Somatochlora arctica 99
Somerset 28, 32, 91, 102, 107, 148, 152, 253, 323–4, 330
Sorbus aucuparia 43
Sparks, Tim 23, 50, 83, 86, 144, 255, **255**, 335–6
Sparrowhawk 182, 200–1
Sparus aurata 228
Spear-moss, Twiggy 59

Special Areas of Conservation (SACs) 318, 326
species abundance *see* abundance
species distributions *see* distribution
Sphecodes niger 109
Sphegina sibirica 110
Spider, Daddy-longlegs 114
    Noble False Widow 113–14
    Wasp 113, **113**
Spider-orchid, Early 208–9, **209**
spiders 75, 77, 113–14, 208
Spinus spinus 212
Sprat, European 223, 291
Sprattus sprattus 223, 291
spring 14–15
    early springs 44–9
Squalius cephalus 283
Squalus acanthias 228
squashbugs 102
Squatina squatina 293
squirrel pox virus 177
Squirrel, Grey 177
    Red 177, 336
St Abb's Head, Berwickshire **215**
St Kilda **223**, 228
State of Nature 310
State of the UK's Birds 151
Steatoda nobilis 113–14
Stenella coeruleoalba 235
Stenobothrus lineatus 105
Stereum subtomentosum 167
Stern, Nicholas **262**, 263
Sterna hirundo 231
    paradisaea 231
Stickleback, Ten-spined 127
    Three-spined 126
sticklebacks 336
Stone-curlew 271, **271**
Stratiotes aloides 277–8
Streptopelia turtur 122
subarctic alpine plants 57–62
Suffolk 320
Sullivan, Charles 259
Sullivan, Charlotte *Paint for the Planet* entry **258**
sulphur dioxide 63, 169

summer 15, 49–51
Surrey 23, 85, 99, 102, 134, 206, **253**
Sus scrofa 265
Sussex 81, 96, 108, 166, 207, 225, 320–1
Swallow 42, 122, 147, 149, 157
Swallowtail **28**, 29, 81, 90
Swan, Whooper 331
Swift 122
Sycamore 43, 50, 69
Sylvia atricapilla 122
    communis 122
    undata 204
Sympetrum fonscolombii 99
    pedomontanum 99
    sanguineum 95
Symphodus ocellatus 240
Syringa vulgaris 43

Tadorna tadorna 317
Taraxacum officinale 54
Taxus baccata 102
temperature changes 13–17
    butterfly emergence 84–5
    Central England Temperature (CET) records 46
    Mean annual temperature in Central England 1950–2015 15
    sea temperatures 215–16
    temperature change and wildlife 17–21
Tench 127
Tephritis cometa 110
    divisa 110
Tern, Arctic 231
    Common 231
Tetrao tetrix 150
Tetrix subulata
Thames 11, 12, 81, 105, 221, 249, 259, 314
Thamnolia vermicularis 60
Thatcher, Margaret 264
Thomas, Chris 90, 150, 255–6, **256**
Thorn, September 93
thrips 208
Thrush, Song 122, 336
thrushes 158

Thursley Common, Surrey 99, 206
Thylacine 30
Thymallus thymallus 127
Thymelicus sylvestris 87
Tick, Sheep **175**, 175–6
tidal power 317–18
Tiger, Garden 91
  Jersey 91
Tilia cordata 18
Timothy 43
Tinca tinca 127
Tit, Blue 122, **123**, 146, 155, 181, 199
  Coal 155
  Great 122, 145, 146, 155, 181, **199**, 199–200
    Egg-laying by Great Tits in Cambridgeshire 1993–2014 *145*
    Synchrony of Great Tit chicks at ten days old and peak of caterpillar production *200*
Toad, Common 130, 132, 134, 177–8
  Midwife 178
  Natterjack 21, 109, 132–3, 203, 218, **218**, 271, 279, 284–5, 320
Toadflax, Ivy-leaved 45
toads 23
  chytridiomycosis 177–8
Topley, Peter 114
Topshell, Flat 220, 221
  Grey 218–19, **219**
  Toothed 220
Tortoiseshell, Large 81–2
  Small 79, 83
Trachurus trachurus 291
Treecreeper 155
trichomonosis 177
Trichophorum germanicum 59
Triggerfish, Grey 228
Tringa totanus 154
Triops cancriformis 203
Tripterygion delaisi 228
Trisopterus esmarkii 228
  luscus 226
Triturus cristatus 30, 122

Troglodytes troglodytes 155
trout 111, 313, 314
Trout, Brown 128, **128**, 186, 282
Tummel River, Scotland 124, 313
Turdus iliacus 122
  merula 122
  philomelos 122
  pilaris 122
  torquatus 158
Turner, J. M. W. 'Fighting Temeraire' 259, **259**
  'Petworth Lake' 259, **259**
Turtle, Leatherback **237**, 237–9
Tussilago farfara 43
Twin-spot Carpet, Dark-barred 93
Tyto alba 155

UK Butterfly Monitoring Scheme (UKBMS) 79, 84, 87
UK Climate Change Programme 264
UK Climate Impacts Programme 36
  MONARCH 271, 279
UK Phenology Network (UKPN) 23, 42, 44, 46, 50, 51, 53, 77, 78, 132
Ulota calvescens 67
University of Kent 133
University of York 90, 255
Uria aalge 230–1
Urospermum picroides 254
Ursus maritimus 335
Urtica dioica 87
Utamphorophora humboldti 101

Vaccinium 191, 192
  myrtillus 270
  vitis-idaea 193
Vanellus vanellus 155
Vanessa atalanta 79
  cardui 80
vascular plants 39, 54, 62, 66, 75, 81, 309
Vendace 129, 281–2, 302
Vernal-grass, Sweet 190
vernalisation 18–19, 49, 198

Verruca stroemia 218
vertebrates 119–20, 160–1
  amphibians 130–5
  birds 144–59
  computer modelling 281–7
  distribution and abundance 123
  freshwater fish 124–9
  mammals 140–3
  phenology 121–2
  reptiles 135–9
  vertebrate experiments 302–3
Vespa crabro 107
Vespula germanica 106
  vulgaris 79
Viola reichenbachiana 197
  riviniana 197
Vipera berus 137
Viscum album 20
volcanic activity 240, 258, 269
Vole, Water 323–4
voltinism 86
Volucella zonaria 110
Volvox 171
Vulpes vulpes 160
Vulpicida pinastri 288

Wales 21, 28, **241**, 312
  amphibians 131
  birds 158, 213
  coastline 221, **237**, 320
  conservation 329
  fish 129, 186, 228, 282
  fungi 167
  insects 109
  lichens 288
  mammals 123
  plants 57, 62, 65, 67, 192
  reptiles 139
Wallasea Island, Essex **332**, 333
Warbler, Cetti's 150, 154, **154**, 212, 331
  Dartford 204, 212
  Reed 201, **201**
  Willow 122, 151
Warnstorfia sarmentosa 59
Wash 320, 324
Wasp, Common 79, 106, **106**
  German 106, **106**

Median 108
Saxon 108
wasps 106, 208
  distribution and abundance 107–8
Water-soldier 277–8, **278**
weather 245
  extreme weather 14, 17, 106, 204, 216, 232, 245–9, 293, 322
Webster, Jon 136
Weever Fish, Lesser 228
Weil's disease 175
Whale, Beluga 226, 236
  Blainville's Beaked 235
  Bowhead 236
  Cuvier's Beaked 235
  Dwarf Sperm 235
Wheatear 122
White, Green-veined 78
  Large **23**, 29, 83
  Marbled 87, 90, 301–2, 328
  Small 78, 83
White, Gilbert 238, 255
  The Natural History of Selborne 82, 107, 152, 158, 206
Whitethroat 122
Whitlowgrass, Hoary 62
Wigeon 156
wildflower meadows 63–4, 106, 265

wildlife 17–21
  future prospects 334–6
Wildlife Trusts 263
Willis, Stephen 90
Willow, Creeping 198
  Snow-bed (Dwarf Willow) 60, 190, 192, 274–5, 298–9
Wiltshire 165
wind turbines 314–16, **315**
Windermere, Cumbria 124–5, **125**, 127, 171–2, 186
winter 15, 19–21, 25, 31, 48, 60–1, 147–8, 155–6, 245, 246–8
  algae 171
  birds 147–8, 155–6, 221
  Bracken 197, 204
  fish 184–5, 225–6, 228–9, 284, 292, 302
  fungi 165–6, 168
  mammals 160–1
  tick survival 175–6
Wolf, Grey 210
Wolffish 228
Wood, Speckled 79, 87, 90–1, 256
Wood-moss, Glittering 190
Wood-sorrel 197
Woodchester Mansion, Gloucestershire 140
Woodland Trust 23, 255

woodlands 195–201
woodlice 75
Woolmer Forest 85, 206
working outside 250–5
Worldwide Fund for Nature (WWF) 263
worms 75, 77
Wrasse, Ocellated 240–1, 243
Wren **154**, 155
Wryneck 271
Wye 128
Wytham Woods, Oxfordshire 143, 199

Xanthorhoe ferrugata 93
  fluctuata 86
Xylocopa violacea 109

Yellow, Clouded 90
Yellow-sedge, Common **59**
  Long-stalked 190
Yew 102, 196
Yorkshire 87, 103, 107, 168, **253**, 296
Yorkshire-fog 43
Younger Dryas 269

Zander 126
Ziphius cavirostris 235
Zootoca vivipara 137